CASE STUDIES IN
CULTURAL ANTHROPOLOGY

GENERAL EDITORS
George and Louise Spindler
STANFORD UNIVERSITY

THE HUTTERITES IN NORTH AMERICA

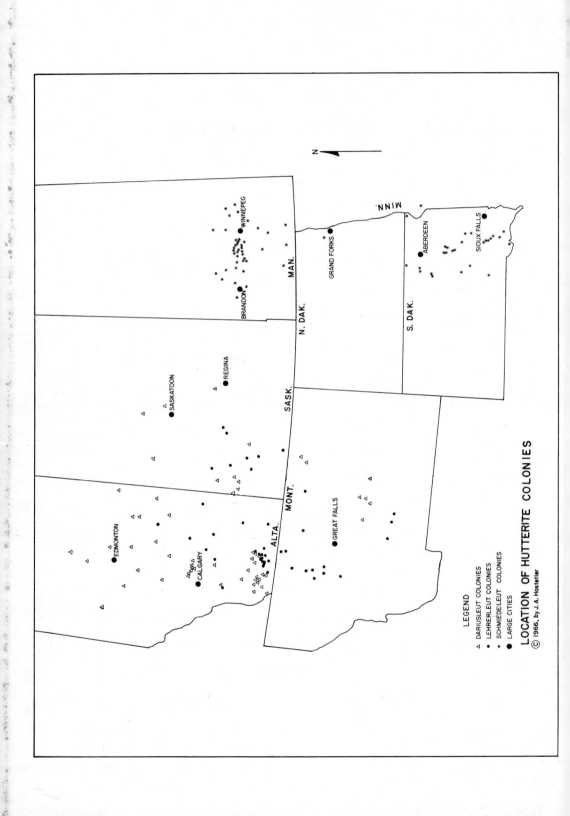

LEGEND

△ DARIUSLEUT COLONIES
△ LEHRERLEUT COLONIES
* SCHMIEDELEUT COLONIES
● LARGE CITIES

LOCATION OF HUTTERITE COLONIES

© 1966, by J. A. Hostetler

THE HUTTERITES
IN NORTH AMERICA

By

JOHN A. HOSTETLER

and

GERTRUDE ENDERS HUNTINGTON

Photographs by Kryn Taconis

CASE STUDIES IN CULTURAL ANTHROPOLOGY

HOLT, RINEHART AND WINSTON

| NEW YORK | CHICAGO | SAN FRANCISCO | ATLANTA |
| DALLAS | MONTREAL | TORONTO | LONDON |

Acknowledgement

*A grant from the United States Office of Education made possible an extensive
study of socialization in Hutterite society, from June 1, 1962 to September 30.
1965, through The Pennsylvania State University as sponsor. Special recognition is
given to Laura Thompson and Calvin Redekop for the preparation of field guides,
to Irene Bishop and Abbie C. Enders for fieldwork, and to Joseph Britton, Robert
Friedmann, Dale Harris, Bert Kaplan, and Karl Peter for substantial consultations.*

Illustration on cover: A Hutterite girl.

Foreword

About the Series

These case studies in cultural anthropology are designed to bring to students in the social sciences insights into the richness and complexity of human life as it is lived in different ways and in different places. They are written by men and women who have lived in the societies they write about, and who are professionally trained as observers and interpreters of human behavior. The authors are also teachers, and in writing their books they have kept the students who will read them foremost in their minds. It is our belief that when an understanding of a way of life very different from one's own is gained, abstractions and generalizations about social structure, cultural values, subsistence techniques, and other universal categories of human social behavior become meaningful.

About the Authors

John A. Hostetler, born of Amish parentage in Pennsylvania, is Professor of Anthropology and Sociology at Temple University. He holds a Ph.D. from The Pennsylvania State University. He has done fieldwork in rural communities in the United States and Canada, and in Switzerland and southern Germany as a Fulbright Scholar, and has taught at the University of Alberta. He has written several books including *Amish Society* (1963) and professional articles dealing with social anthropology and problems of marginal persons.

Gertrude Enders Huntington (Mrs. David C. Huntington), an anthropologist, lives in Ann Arbor, Michigan, where her husband teaches at the University of Michigan. They have three children. She attended Oberlin College and was graduated from Swarthmore College, attended Rochester University, and holds a Ph.D. degree from Yale University. She has taught biology, done fieldwork in Turkey, and wrote a five-volume dissertation entitled "Dove at the Window: A Study of an Old Order Amish Community in Ohio." She served as principal fieldworker on the Hutterite Socialization Study, of which John A. Hostetler was project director. She and her family lived as Hutterites in a colony for an extended period.

About the Book

The authors describe in intimate detail the day-to-day living patterns of the Hutterites, a communal group living in the Great Plains of the United States and Canada. This personalistic treatment of the materials allows the reader to empathize fully with the Hutterites as persons.

As a group, Hutterites are protected from the outside world by a well-organized belief system, which offers a solution to their every need. The Hutterite community minimizes aggression and dissention of any kind. Colony members strive to lose their self-identity by surrendering themselves to the "communal will," and attempt to live each day in preparation for death, and hopefully heaven. It is not surprising, therefore, that the Hutterites have retained their solidarity despite the persecution of surrounding neighbors.

The authors—careful students of Amish society, a society resembling the Hutterites in many aspects—view the principle of order as a key concept underlying the Hutterite way of life. Order is synonymous with eternity and godliness; even the orientation of colony buildings conforms to directions measured with the precision of a compass. There is a proper order for every activity, and time is neatly divided into the sacred and the secular. In the divine hierarchy of the community, each individual member has a place—male over female, husband over wife, older over younger, and parent over child. It is also noteworthy that the Hutterites find the rigidity of German syntax (which is applied whenever possible) supportive of their hierarchical systematizing.

One of the first questions an interested outsider might ask when viewing the well-articulated system is: Why does it work? The authors' answer would be, "Hutterite society is a school, and the school is a society." The Hutterites do not value education as a means toward self-improvement but as a means for "planting" in children "the knowledge and the fear of God."

The authors take the child from birth to death, describing in detail what happens to him at each stage of his socialization and enculturation—how, as soon as possible, he is weaned away from his parents into the group where the setting minimizes his treatment as an individual and maximizes his identity as a member of the group.

Although the history of the Hutterites is a history of persecution, it is relatively easy for the well-trained Hutterite adult who has internalized the God-given rules to withstand persecution. The harassment, humiliation, ridicule, and torture of this people has served to reinforce the basic Hutterite belief that there is only God upon whom to depend.

GEORGE AND LOUISE SPINDLER
General Editors

Stanford, California
July 1967

Contents

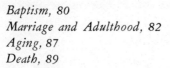

Introduction

THE PEOPLE DESCRIBED in this book are attempting to establish a colony of heaven. They do not have illusory ideas that their colony is perfect, but they have acquired some utopian-like characteristics in their social patterns: economy of human effort, elimination of extremely poor or wealthy members, a system of distribution that minimizes privileged position, motivation without the incentive of private gain, and a high degree of security for the individual. All of these characteristics are contained in a communal society without the use of a police force and with an ideology that reasonably satisfies both spiritual and material needs.

By living on large acreages of communally owned land, the Hutterites maintain a degree of geographic isolation. Their German dialect and distinctive dress reinforce social isolation from their neighbors. The Hutterites think of themselves as a people who honor God properly by living communally. Honoring God, they say, requires communal living, devout pacifism, and proper observance of religious practices. They regard as inevitable the atrocities they have suffered throughout their history, as well as the misunderstandings with the outside world in modern times. To the Hutterites, suffering is a necessity for which they must be morally prepared at all times. During four centuries, the group has demonstrated a remarkable ability to adapt to changing political, social, and technological environments. The Hutterites are the largest family-type communal group in the Western world. The society is noted for successful large-scale farming, large families, and effective training of the young. These distinguishing features are inseparable from their religion. All Hutterites inhabit the United States and Canada and none have survived in the countries of their origin, namely, Austrian Tyrol and Moravia. In 1965 the population numbered about 16,500 persons in 170 colonies.

The Hutterites originated during the Protestant Reformation in the sixteenth century and are one of three surviving Anabaptist groups. The other two are the Mennonites (Smith 1957) and the Swiss Anabaptists including the Old Order Amish (Hostetler 1963). The Anabaptists were nonconformist groups who rejected infant baptism and membership in a state or all-inclusive church. Instead of reforming the medieval church, the Anabaptists wanted to withdraw from its influences entirely and found voluntary church groups. Their views challenged existing social, economic, and political institutions as well as those of the great re-

1

formers including Martin Luther. Many became martyrs. With the collapse of feu-
dalism of the Middle Ages, there were many landless persons—artisans and craftsmen along with scholars—who joined newly formed movements such as the Anabaptists. They were the less privileged of their day, and the nonconformist movements offered a new psychological outlook on life. Equipped with the justification
for denying earlier traditions, they substituted small voluntary groups for the traditional social hierarchy of feudal society. Various Anabaptists wished to establish a
Christian-type community in which private property would be abolished and temporal possessions would be surrendered voluntarily by the individual (Klassen 1964).
Of many attempts at forming "communities of love" (Troeltsch 1931:331–334)
only the Hutterites successfully and permanently carried out communal living. In
their early period Hutterites were a "conversionist" type of sect, but in their more
recent history they approximate the "introversionist" type (Wilson 1959), avoiding
involvement with the larger society.

The Hutterites regard 1528 as their founding date (Zieglschmid 1943:85).
While fleeing from Nikolsburg to Austerlitz, a group of religious refugees introduced the practice of "community of goods." Each person heaped all of his possessions on cloaks that had been spread on the ground, and stewards were selected as
overseers of the material goods. Thus, occasioned by necessity and sanctioned by religion, communal sharing was made the norm. The first common household (*Bruderhof* or colony) was founded in Austerlitz in Moravia. Jacob Hutter of Tyrol
joined the community and greatly intensified the discipline. He was captured by the
authorities, interrogated, brutally whipped, and burned at the stake in 1536. Although not the founder of the Hutterites who are named after him, Hutter was
their most outstanding organizer and leader (Horsch 1931).

The Hutterites were tolerated for a time by local barons of Moravia who
found them to be outstanding farmers and craftsmen. During this time of tolerance
and protection they experienced growth and prosperity. They expanded to approximately eighty colonies with an estimated population of about 20,000 persons
(Rideman 1950:275). Their colonies were large and well managed. Their activities
in addition to agriculture included ceramics, cutlery, milling, wine making,
spinning, weaving, tailoring, clock making, and carriage making. Their mineral
baths achieved a wide reputation. Hutterite nurses, midwives, and physicians were
in great demand. All young men learned a trade. The schools were systematic, com-
pulsory for children of members, and were attended by non-Hutterites. Virtually all
of the members were literate.

Prosperity was followed by growing intolerance. The enemies of the Hutterites were not only religious and political, but soon there developed economic envy
and antagonisms. They were accused of monopoly. The Hutterite workers were
thrifty, motivated, and efficient, and myths of Hutterite wealth spread widely, which
later induced raids and robbery. Their pacifistic communism, disinterest in politics,
and refusal to fight on any side in a war, rendered their future precarious.

With the outbreak of war between Austria and Turkey in 1593 the Hutterites were raided by both Austrian and Turkish armies. At first they were heavily
taxed, then robbed, plundered, tortured, and many were executed or sent into cap-
tivity. The disasterous Thirty Years' War followed, and by 1622 all the Hutterites

had been driven from Moravia. Slovakia and several small states to the east, including Transylvania and Wallachia, became host to the group. In Slovakia they rebuilt their communities only to suffer decline, steady pressures from the Jesuits to return to the Catholic faith, and imprisonment. With the elders in prison or in monasteries and the children in orphanages, many accepted Catholicism. A remnant crossed the Carpathian mountains into Wallachia in 1767 where they were soon caught in the Russo-Turkish War. On invitation from a Russian general, 123 Hutterites settled in the Ukraine in 1770.

Although the Hutterites were granted complete religious freedom, exemption from military duty, vast agricultural lands, and free practice of their trades, they were wary of losing their privileges in Russia. Upon achieving reasonable material recovery after so many years of hardship, their best craftsmen refused to turn over their profits to the common treasury. Communal ownership was abandoned for forty years, from 1819–1859. A period of internal decline, resulting in poverty and illiteracy, was followed by renewal. An extensive religious revival occurred in 1856–1859. Scarcely had the group reinstituted communal living when the Russian government changed its attitude toward German colonists. Expert farmers and settlers were no longer needed. With large numbers of unassimilated Germans living on choice lands in the Ukraine, a new nationalization policy was enacted that forced the Hutterites along with many Mennonites, Doukhobours, and other minorities to migrate to North America. The universal military training act of 1872, repealing earlier exemptions including those granted by Catherine the Great, precipitated their decision to leave the continent.

All Hutterites, numbering nearly 800 persons, relocated in South Dakota. Their need for large isolated blocks of land rendered them ineligible for homesteading privileges. About half of the families decided to take advantage of homesteading by settling on family farms. These *Prairieleut* (prairie people or "noncolony" people), as they were called by the Hutterites, later affiliated with nearby Mennonite groups. The Hutterites founded three colonies in South Dakota from 1874–1877, each immigrating at slightly different times. Hutterites today acknowledge three distinct "people" (*Leut*) among them. Taking their names from their first leaders in the United States, they are *Schmiedeleut, Dariusleut,* and *Lehrerleut*. The three share a common body of doctrine, language, and social patterns, but each has its own senior elder and *Ordnungen* (discipline). Each *Leut* has its own periodic preacher assembly that ordains leaders and modifies or changes the discipline, and each is unified by the preferred endogamous marriage pattern. The *Leut* is the largest unit within which there is both a means and a moral obligation to settle disputes.

The research on which this case study is based was part of a study of communal education in Hutterite society over a period of several years (Hostetler 1965). The units of observation consisted of one colony from each of the three *Leut* and are anonymously referred to in the case study as *Dariushof, Lehrerhof,* and *Schmiedehof,* though they are not necessarily representative of other colonies within the *Leut*. The three colonies observed include respectively a newly formed small colony, an average-sized colony, and a large colony that has had one branching. One is located in the United States and two are in Canada. The three are no closer than

500 miles from each other. The criteria for selecting the three colonies took into account size, age, social and economic practices, and *Leut* affiliation. At the time of this study, the populations of the three colonies were 55, 101, and 124, respectively, or a combined population of 280 persons in thirty-eight nuclear families. During the course of the research, 110 additional colonies were visited. The objective was to observe a small number of people intensively in order to gain a detailed record of interpersonal relations in each of the sample colonies. The data was obtained by direct observation and from conversation while participating in the culture. Information schedules or questionnaires were not used.

Throughout the case study English equivalents will be given for the German terms peculiar to the culture except when special emphasis is desired. Of the several possible names for the group, such as Hutterian Brethren, *Huterische Brüder, Huterischen,* or *Taufergemeinschaften,* it was decided to use the more familiar term Hutterites. It is the name established by usage by the group members as well as by others.

1

World View

DAWN BREAKS on the vast fields of the prairie, lighting a cluster of buildings hidden in the quiet recesses of a coulee. On a summer morning the colony bell wakes everyone at 6:15 A.M. "Like everyone else," says a Hutterite girl, "I roll out of bed and get dressed, then go down on my knees to thank the good Lord for protection and to get a blessing for the day."

The bell calls the adults to the communal kitchen for a breakfast of prunes, cheese, smoked ham, bread, jam, and coffee. Quickly and in order the men file in, hang their hats on a rack extended from the ceiling, and take their assigned places on benches around a long table. All are dressed in black work trousers and colored shirts, and all of the mature men are bearded. Uniformly garbed women with identical hair styles enter next, wearing polka-dotted head scarfs, long, patterned aprons and dresses. They sit at a second long table separate from the men. The last are scarcely seated when a man gives the audible signal to pray. Instantly everyone raises folded hands and a man mutters a short prayer.

The table tops shine with new varnish. Each person uses a shallow dish for his individual plate. The bread, on wooden centerpieces, is within reach of each set of four persons. Coffee is poured into cups by the oldest person, who is always seated at the end of the table, and is then passed to each person. No children or babies are present; they have been left in the apartments and will eat separately. The highest officials of the colony, the preacher and his assistant, eat in the preacher's apartment where food has been brought to them from the central kitchen.

Silence prevails during mealtime. Everyone eats quickly, and breakfast is dismissed with another prayer. Each person carries his own dishes to the kitchen. Women quickly finish clearing the tables and begin washing the dishes. In seven minutes the job is completed.

The bell rings again, this time signaling the children from ages six to fourteen to their dining room. Those from two and one-half to five years go to the kindergarten for their breakfast and will remain there for most of the day. The men gather in the shop, where every day begins with each person receiving an assigned responsibility.

Anthropologists who thoroughly familiarize themselves with a culture can formulate the underlying principles that govern behavior patterns. They assume that the patterns of a culture are not random and haphazard but represent a selection based on deep-lying assumptions. All peoples have norms of belief and norms of practice. The beliefs that a people express about the nature of the external world, about man himself, and about the nature of existence have been called existential postulates (Hoebel 1966:23). While these ideals direct the behavior of a society, they are never wholly effective in solving the problems of day-to-day living. Another set of assumptions called normative postulates specify whether behavior is good or bad, proper or improper. The existential and normative postulates (also called ideal versus real, or formal versus informal culture patterns) provide individuals in a society with a distinctive orientation toward the world around them and toward each other. This distinctive approach to life, which colors the entire view of things, is known as *Weltanschauung,* or world view. Thus, the rules of living seem natural and right to individuals who have absorbed the world view of their culture. To outsiders who are not of that society they seem strange and unnatural.

The Hutterites draw their ideals from the Hebrew-Christian Bible, the Old and New Testaments, and from the Apocrypha, but their solutions to day-to-day living differ so sharply from most Christian groups that they scarcely bear resemblance to them. The official statement of their faith was written in prison in 1540 by one of their leaders, Peter Rideman (1950). It embodies the Hutterite view of life and death, relation of man to man, and the basis for communal living. Formulated in this chapter are the existential postulates of the Hutterites, based on the Rideman book and on observations in contemporary colonies.

View of the Universe, Nature, and Supernature

Absolute authority emanates from a single supernatural being, an omnipotent God, who created the universe and placed everything in order and harmony. No man can cause events to happen, and all events that affect a person's life are determined by God. Man was made to worship God, the Creator, and not to worship the creation or things made by God. Orientation is toward life after death (eternity) and not toward the enjoyment of the present life nor toward self-development of the individual.

In their morning and evening prayers, children acknowledge an "eternal God who has wonderfully created all things in Heaven and on earth." In catechism at school they learn the difference between a "right faith" and a "false hope." A right faith is both "living" and "a gift from God," made possible by "believing in God's word," and demonstrated by "pious living" and suffering. When baptized at the age of about twenty, the young adults "establish a covenant with God and all his people" to "give self, soul, and body, with all possessions to the Lord in Heaven." In marriage a spouse is acknowledged as "a gift from God."

Hutterites consider themselves responsible for the maintenance of their own offspring through vigilant daily teaching. "Through attentive hearing and observing of the gospel we become partakers of the community of Christ" (Rideman

1950:43). Maintenance implies not only nurture but vigorous separation from the world. Diligent nurture is required at all times for all ages, from the youngest to the oldest, since "no man is free from the tendency to serve his natural inclinations."

Respect for order and authority pervades Hutterite thought and practice. The natural, including human nature, must obey the supernatural order. The right order regulates all things, the moon and stars, the plants and animals, and also man. Just as day is separated from the night, so God established order among human beings. The right or divine order requires a hierarchy of relationships. One part is always superior to the other. The higher cares for, directs, and uses the lower; the lower serves and obeys the higher. God is Lord over man, man is over woman, and the elder over the younger. Man has power over the animals. Man is ruler over material things, inventions and machines, and he may use them as long as the proper relationship and function are observed. Man may not change the order of God. He may not, for example, kill other men, nor may he interfere with the process of natural conception.

The reason for the creation and continuation of the universe, according to the Hutterites, is that man might honor God. Only those who "submit and surrender" to God are His children (Rideman 1950:68), and submission is evidenced by forsaking everything, including property, in order to live communally. God desires to be honored only in His Church and by none other, and the Hutterites believe they are unique in using "created things" to lead men to God in such a narrow path. Their way of life justifies the existence of man and of the material world that was created for man's use.

Although sixteenth-century Hutterites were very aggressive in missionary activity (Littell 1958:126), contemporary Hutterites acknowledge their inability to send out missionaries but apply the missionary directive to vigorous teaching of the young. "Weeding the garden at home," is believed by some colonies to be the first responsibility and, by others, their sole responsibility. Hutterites do not consider themselves responsible for organized missionary work, but when a believing individual approaches the colony, efforts are taken to show the penitent one the correct way.

View of Human Nature and the Outside World

Human nature is regarded as evil, fallen, and displeasing to God. Man was made in the image of God, but through the disobedience of the woman in the Garden of Eden, Adam and Eve and their offspring became perverse. Man fell from spiritual nature to a carnal nature, and children are born in sin and with a tendency to desire the carnal.

The individual can be restored to the spiritual nature by believing in the Word of God as interpreted by Hutterite sermons, by repentance of sin (culturally defined misbehavior), and by continuous and daily surrender of self to the will of God in communal living. By thus "dying" to his carnal nature, man can be "born of Christ" and receive a spiritual nature, a nature that will live eternally.

Before the individual is old enough to acknowledge voluntarily his own sinful nature, he is taught to obey God and revere Him as the "Almighty Father." Children are not permitted to carry out their own headstrong and carnal practice, and from the beginning are taught "the divine discipline" (Rideman 1950:130). "Just as iron tends to rust, and soil nourishes weeds if not cultivated, so children require continuous care to keep them from their natural inclinations toward injustices, desires, and lusts" says a seventeenth-century elder, Andreas Ehrenpreis. Baptism is postponed until the threshold of adulthood when full implications of the denial of private property and the acceptance of suffering are understood.

Hutterites view the world dualistically. The carnal nature is temporal and passing and brings death to man. By contrast, all who are "born of Christ" are ruled by the spiritual nature, which is eternal. (See Eliade 1959:197, for the recurring theme of the second birth in other religions.) The two kingdoms are separate and each must go its separate way. It follows that Hutterites aim to live separate from the (carnal) world with their loyalties rooted in the spiritual. Separation is ordained by God, says Rideman (1950:112, 139, 140, 187). "Many tribes call themselves Christian but they do not wish for the Kingdom of God and Christ. The Kingdom of God is the cross, tribulation, suffering, and persecution, to drink the bitter wine of suffering and to help Him carry His cross. They [other professing Christians] neglect such suffering. They do not like to be hated by the world" (Rideman 1950:3).

The cleavage between the carnal and the spiritual world, between Hutterites and outsiders, is vividly maintained on the informal level. Their loyalty has been tested by scorn, persecution, and martyrdom throughout their history. The many accounts of persecution in their chronicles (Wolkan, Zieglschmid) are taught to each new generation to impress upon the young the price of absolute loyalty to the beliefs. Two accounts, one from the sixteenth and one from the twentieth century, will illustrate the concreteness of the cleavage.

The first illustration is from an imprisoned Hutterite who said: "My shirt decayed on my body. There was not a single thread left of it except for the collar. I did not see the sun for one and a half years. In the dark and awful dungeon I no longer knew the difference between day and night. All of my clothes rotted so that I was stark naked. The insects and the worms ate my food as soon as they smelled it."

The writer of these lines was Hans Kräl, a Hutterite preacher, who in 1557 was seized in Tyrol. Placed in the stocks for thirty-seven weeks, his interrogators periodically had him pulled up from the dungeon with a rope to persuade him to give up his faith. The light hurt his eyes so much that he was relieved when they lowered him again into the dark, vermin-infested dungeon. After twenty-three months of imprisonment he received an edict from the government: "Because you are obdurate and will not accept good advice, you are sent to the sea as a galley slave." The jailkeeper said, "Only hard labor and whippings are in store for you there." But Kräl answered, "I will trust in God my Master who is present on the sea as well as on the land, who gives me aid and patience."

Kräl nearly perished in filth and vermin. Many of his contemporaries died, and others lost their limbs as a result of imprisonment in the stocks. Mice and insects carried away their decomposed toes and fingers. Later in his life Král escaped from the galleys and became a leader of the Moravian colonies. He composed a song based on his dungeon experiences that is still sung today in all colonies (*Die Lieder...* 1962:538).

Persecution from the carnal world was not confined to unsympathetic Europe or limited to the sixteenth century. In 1918, a group of young Hutterites in South Dakota were summoned for army duty. They obeyed orders to appear at Camp Lewis, but there they refused to sign admission papers, to put on army uniforms, or to take up any kind of duty, on grounds that they were religious objectors to war. After two months in the guard house, four young Hutterite men were sentenced to thirty-seven years in prison. They were taken to the notorious military prison on the island of Alcatraz in San Francisco Bay. Four armed lieutenants attended the young pacifists who were handcuffed during the day and chained by the ankles to each other by night. At Alcatraz they were ordered to remove their outer clothing and put on military uniforms, but they refused. They were taken to the "dungeon" of darkness, filth, and stench where they were placed in solitary confinement. The guard placed a uniform in each cell and said, "There you will stay until you give up the ghost—just like the last four that we carried out yesterday."

For several days the young men slept on the cold, wet concrete floor wearing nothing but their light underwear. They received a half glass of water every twenty-four hours but no food. They were placed at distances that made it impossible for them to speak to one another. They were beaten with clubs and, with arms crossed, were tied to the ceiling. After five days they were taken from the "hole" for a short time. Their wrists were so swollen from insect bites and skin eruptions that they could not put on their own jackets. For the remaining months at Alcatraz they were allowed one hour of outdoor exercise each Sunday afternoon.

After four months at Alcatraz the men were transferred to Fort Leavenworth, Kansas, by six armed sergeants. They arrived at their destination at 11 P.M., after four days and five nights of travel, chained together two by two. From the railway station to the military prison they were driven on foot through the streets and prodded with bayonets. Although they were handcuffed they managed to carry their satchels in one hand and their Bibles in the other. On arrival at the prison, soaked with sweat they were compelled to remove their outer clothing. Two hours later, when they received their prison clothing, they were chilled to the bone. In the early morning hours they stood outside and waited in the cold. Joseph and Michael Hofer collapsed and were taken to the hospital. Jacob Wipf and David Hofer were sent to solitary confinement and placed on a starvation diet. They were made to stand nine hours each day with hands tied and stretched through the prison bars, their feet barely touching the floor.

Wipf managed to send a telegram to the wives of the two Hofer men who were hospitalized, and the women boarded the earliest train. The depot agent insisted that the telegram had come from Fort Riley rather than from Fort Leavenworth, and issued tickets to the wrong destination. After losing a day, the women arrived at midnight to find their husbands near death. When they returned in the morning, Joseph was dead. The guards refused Maria permission to see the dead body. She talked to the colonel, pleading in tears to see her husband's body. She was taken to the casket only to find that the body of her husband had been dressed in the military uniform he had so adamantly refused to wear. Michael Hofer died several days later. The wives and a few other relatives accompanied the bodies to their home community, where enormous funerals seared the Hutterite minds with the price of true apostolic faith.

The above accounts illustrate not only the uncompromising loyalty of the Hutterite to his community and to his religious faith, but also the reality of the conflict produced by his dualistic world. The two Hutterites who were faithful to their death, like their contemporaries, were prepared to die. Their culture prepared them to expect hostility in life. Death is a release from a "valley of tears," just as it is a release from a world that has rejected their concept of the good life.

View of the Ego and the Relation of the Sexes

In the Hutterite view, the individual will must be broken. This is achieved early, primarily during the kindergarten age, and is reinforced continually until death. In place of self-fulfillment there must be self-denial. The individual must be humble and submissive. After approximately twenty years of intensive indoctrination, the individual is expected to accept voluntarily the teachings of the colony. When he is able to express the remorse, abasement, and loathing associated with his sinful self, he will receive baptism. Individual identity must be fused with the community. Just as a grain of wheat loses its identity in the making of a loaf of bread (Rideman 1950:86), so the individual must lose his identity in one corporate body. "God worketh only in surrendered men."

Self-surrender, not self-development, is the Hutterite goal. The communal will, not the individual will, becomes important. The good of the majority governs the stages of life from birth to death. Since human nature is sinful from birth, Hutterites value education as a means for "planting" in children "the knowledge and the fear of God" not for self-improvement. The consequences of original sin are moderated by intensive teaching from an early age. "We should let the Heavenly Gardener implant such fruit that will bear everlasting life. The child's will should not be fulfilled, but the Father's will. Our will should be locked in His will."

Although the individual is "grafted into the divine character or nature" (Rideman 1950:18), he is never free from temptation of his carnal nature. His security in the divine order depends not on his verbal affirmation or the rite of baptism, but on his "proven works" or daily behavior. Behaviorally this means adherence to the rules of the church community. The individual may not conform to his own interpretation of the "word" or his own notion of obedience. Since "God worketh in surrendered men," the individual must submit to the will of the community because community is the will of God.

The individual must never pray "My" Father but "Our" Father. Thus, one prays for the common good of the majority and not for the good of the individual. Since the community is the body of Christ, God dwells in the community.

The individual has a proper, defined place in the community. By divine order male is over the female, husband over wife, older over younger, and parent over child. Women have neither vote nor a passive participation in the formal decision. Marital relationships proceed from the divine order. God rules over the soul or spirit, the spirit rules over the body, and man rules over woman. As a woman should obey the man, so also the body should obey the spirit. In creation man has lordship over woman while the woman has "weakness, humility, and submission." Woman should "heed her husband . . . and do all things with and naught without his counsel." If she does not, she forsakes the order of God. Man "should have compassion on the woman as the weaker instrument" and must care for her in temporal and spiritual things.

The purpose of marriage is to instruct and lead men to God. If the marriage is not rightly regarded and observed, "it leadeth men away from God and bringeth death." A husband must be head of only one wife as Christ is the head of only one church. A marital union can be broken by transgression of the marriage promise, and if the husband "as the glory of God" dishonors himself through the weakness of the flesh, he breaks the union. A broken promise does not lead to divorce; reconciliation is possible only through repentence and restoration of the broken covenant. To achieve marriage "one should ask not his flesh but the elders that God might show" through the community "what God has provided from the beginning." Man and woman come together not through their own action and choice, but in accordance with God's will and order, and in fact, obedience to God and community are above obedience to spouse. Thus if one partner forsakes the faith, the other remains bound to the community.

Physical attractiveness is a secondary consideration in marriage, while the primary factors are those which benefit the colony as a whole, namely, whether the girl is a good housekeeper, likes children, is obedient to the colony, and whether the boy shows initiative in his assigned work and accepts the values of the colony.

View of Community and of Property

Living communally is believed to be the divine order of God, who from the beginning created all things for common use. Through private possession, it was man who brought disorder into the world by his grasping and greedy spirit. The dichotomy between the carnal and the spiritual is expressed in the right use of material goods. Carnal men are viewed as living in perpetual covetousness by making property, food, land, and "created things" the object of private gain. The sun, moon, day, and air, and material things are for the good of all men, and to make them private property is to ignore God's order. "Therefore, whosoever will cleave to Christ and follow him must forsake such things." Just as the Father and the Son in Heaven live in common, and hide nothing from each other, so his children should also live (Rideman 1950:88–90).

God's "gifts are not given to one member alone, but for the whole body with its members." "The communion of saints must show itself not only in the spiritual but in temporal things also." "Through such wrong taking and collecting of created things he [man] hath been led so far from God that he hath even forgotten the Creator." Privately owned material possessions lead human beings away from God. In his carnal state of mind man honors "created things" instead of making them subject to him (Rideman 1950:88).

Honest manual labor, making what is useful "for the benefit and daily use of man" is honorable, but merchandising (buying and selling for profit, without the contribution of labor) is "a sinful business" (Rideman 1950:112, 127). Material goods, as well as capital earnings, are for the welfare and benefit of the community. Every member "shall give and devote all his or her time, labor, services, earnings, and energies to . . . the community, freely, voluntarily, and without compensation. . . ." (Hutterian Brethren Church, Constitution, 1950). The individual, in turn, together with his dependents, are "supported, instructed, and educated" by the community.

The community is more important than the individual and governs the activity of the individual, and the corporate group has the power to exclude and to punish, to forgive and to readmit. The German teacher is charged with the formal instruction of the young and he is their primary disciplinarian. The father, as head of a family, supports the discipline of the colony, and the colony can require him to punish his own child in the presence of the group. The older person is required to correct the younger, regardless of his relation to the offender, and self-assertion by the individual against the group is not permitted. The duties of each person are assigned by the community, and tasks may not be chosen or positions aspired to by the individual. Disobedience to the community means forsaking the commandment of God, and sin must be punished in proportion to its severity. Unconfessed sins will be held against the individual on the day of judgment and punishment will be meted out in the afterlife.

Language and Reality

Hutterites view the world through Germanic-language thought patterns and from a perspective common to medieval Europe before the dawn of modern science. For them the word of God is contained in the German language, and since the word of God is believed to be eternal, the language forms and style of thinking convey symbolic spiritual meaning. Due to limited contact with other language groups, including other German-speaking people, and isolation from changes in intellectual thought, a type of pre-Lutheran language pattern and logic permeates their social patterns. The lack of interest in changing the destiny of history and of controlling nature through critical and scientific logic are distinctive of their view of the world.

Hutterites are trilingual, but have limited ability in three languages. The mother tongue of all Hutterites is a household dialect similar to that still spoken in certain parts of Austria, South Tyrol, and Bavaria. The child learns limited High German from biblical and religious sources in the German school. A speaking knowledge of English is acquired by all Hutterites in the English school. German and English usage are symbolic of the two kingdoms, the colony and the world. High German is used for all sacred and ceremonial occasions, and the sermons read in church have been copied from seventeenth-century German texts. "We believe," said a preacher, "that our faith can be expressed more deeply, sharply, and fully in the German language."

The rigid character of the German sentence structure is supportive of the absolutism of the authority patterns. Language binds the thoughts of its speakers by the patterns of its grammar. For the Hutterites the supremacy of the German language is illustrated by their sermons. The words "scripture" and "sermon" are interchangeable in Hutterite usage. When a Hutterite preacher wants to know the interpretation of a biblical passage, he will rely on his old sermon books. Sermons are the unchallenged source of authority, and no individual, not even a preacher, may interpret the Bible without the aid of the traditional sermons. They say "No one among us is spiritual enough today to preach without the written sermons; it would be very unsafe." Preachers do not embellish their sermons with contemporary illustrations, as this would "distract us from the word of God." This dedication to the Gothic German handwritten sermons insulates the ideology from competing outside influences. Critical scholarship is considered unnecessary, for such scientific-individualistic conceptions militate against the wisdom of consensus.

When several Hutterite boys were serving jail sentences as conscientious objectors during World War II, the authorities forbade them the use of their German books. Friends volunteered to translate their hymns and sermons into English, but when the elders examined the translations they declared that the meaning had been so watered down as to be worthless. The identification with the German language is so strong that the traditional Gothic printed type is equated with the sacred. The hymnal and school books, when reprinted by outside firms, must remain in Gothic typeface. Hutterite handwriting, too, employs the script alphabet, and not the so-called "English" (Latin) alphabet now commonly used in modern Germany. The unchanged form of words, phrases, and sentences is an example of the power of Holy Writ. Parallels in other cultures are Hebrew of the Old Testament, Arabic of the Koran, Latin in the Roman Catholic Mass, and Zuni ritual poetry when memorized "word perfect" and unchanged.

Private interpretation of the word of God is forbidden. No individual Hutterite, no matter how mystical or saintly his behavior, has legitimate authority to interpret the word of God. For the Hutterites there is no "inner light" to challenge either the word of God or the religious tradition of the community. The decaying Bibles and sermon books are placed in the casket of a deceased person. Here the pages of the worn-out book and the body decay, but the word of God and the soul are identified with the eternal and live forever.

Verbal sharing is very important to the colony and supportive of the communal social organization. Although the Hutterites are literate and capable of writing in both German and limited English, literacy is less important to them than verbal sharing. Oral sharing of words and phrases in ceremonial practices lends strong support to uniformity of thought patterns and communitarian values. In all colony schools from kindergarten through the rite of baptism, words and phrases are memorized and repeated in unison. All memorize the same or similar prayers, the same hymns, and the same vows. The right hymns must be sung in the right way "for the glory of God"; they must not be sung in a manner to make a "sweet sound" or for "carnal joy." Ritualization of prayers before and after meals, before retiring and when rising, are always considered necessary. A Hutterite is so thoroughly conditioned to ritual at mealtime and at coffee time that his hands form a prayer posture spontaneously before and after eating, whether the place is the dining hall, his apartment, or a roadside restaurant. Thus access to God, to food, to sleep, and to a new day is found through stylized prayer recited in the proper way. Individual fulfillment is achieved by fusion with the group in daily recitation and singing and in the evening church service. The blending of words and phrases with the communal whole allows little place for individual assertion, and individual utterances are undistinguishable from the whole. In verbal sharing, as in property ownership, the individual's voice is blended and lost with that of the group.

In a society that permits no material object to become sacred, special importance is attached to words, language patterns, and oral tradition; there are virtually no religious relics. Ritual centers around words, not objects or property. Bibles and sermon books are the most valued property, but the contents (the word) and not the material itself is believed to be holy. Even in baptism the word of God and not the water is holy. Passages from sacred writings are more easily shared than material objects, group recitation and singing being the most common forms of sharing. So intensive is the sharing that persons in the church service do not possess individual hymn books. Only the preacher has a book, and after reciting each line the group sings in unison.

The Hutterites have retained their medieval speech patterns for over four centuries even though they have borrowed expressions and words from countries in which they have lived. An analysis of their speech (Obernberger 1966) reveals a similarity to the languages of contemporary southern Bavaria and Austria, particularly the area of Carinthia, Austria. The southern Bavarian dialect is also spoken in parts of Tyrol, and a sprinkling of vocabulary reflects their stay in Slavic areas, Transylvania (Rumania), the Ukraine, and the United States. The terms for in-law relationships (*Schweigermutter, sohn,* and so forth) reflect the pattern of Carinthia. The term for cucumber (*Kratsewets*) apparently entered the Hutterite speech pattern from Transylvania. From the Slavic languages there is *dšainik* for kettle, and from the Ukranian are the words for teapot (*cǎjink*) and cap (*xatǎs*). The many English words used today by Hutterites center around technology and economic habits not common to them before their immigration to this country. English more than any other non-Germanic language has had the greatest influence on their vocabulary. On the whole there is a consistency in the traditional speech pattern with no observable changes in the ceremonial vocabulary. The Hutterites have synthesized a pre-Lutheran speech pattern with their doctrine so as to form a distinctive view of the world. This view has been supportive of their social organization for more than four centuries.

Summary of World View

These existential postulates are the major ones professed by Hutterites. They provide individuals with cognative guides to "reality" and lend structure to the life style of individuals. The manner in which professed beliefs are practiced varies a great deal from one society to another, and even within Hutterite society they vary in detail from one colony to another. The applications of the formal beliefs to the solutions of day-to-day living (normative postulates) will become clear in the chapters that follow.

As a corporate group who in a real sense are a minority culture strikingly different from populations surrounding them, the Hutterites have articulated many reasons for their separateness. The significance of birth, life, and death are determined by the culture, and knowledge of the Hutterite culture enables us to understand why from birth the people of that culture are believed to have the tendency to love carnal, personal pleasures, and how they manage to cope with a dual nature.

During his life span the Hutterite has the opportunity to change his destiny by giving up his individual will for the welfare of the community. This change in destiny requires a lifetime of submission to communal living in order to achieve the hope of eternal life after death. Death is managed in such a way that it becomes not a dreadful experience devoid of purpose, but the final triumphant step in restoration to a divine order.

The Hutterite world view includes the following:

1. Absolute authority resides in a single supernatural being, an omnipotent God, who created the universe and placed everything in a divine order and hierarchy. All events are ordered of God and nothing happens without the knowledge of God.

2. Through transgression of the divine order and the disobedience of man toward God, all nature became perverse. In its fallen or carnal state human nature desires self-recognition, self-ownership, and bodily (carnal) pleasure. Human nature is helpless and can never completely overcome the carnal tendency, and only by believing the word of God, by repentance, by receiving the grace of Christ, and through continual submission of the self to the will of God in communal living can the individual attain eternal life after death.

3. The carnal nature and the spiritual nature are inevitably antagonistic to each other and constitute two separate kingdoms. The fallen or carnal nature is displeasing to God; the spiritual nature permits an individual to be restored to the divine order where he will voluntarily share his life and his possessions.

4. The spiritual kingdom is ruled by the spirit of Christ and is known by the complete obedience experienced in sharing all material goods. Only in a divinely created fellowship, separated from the world, can men succeed in living communally, and only in this way can God be properly honored, worshipped, and obeyed.

5. At all times the individual must be submissive to the will of God that is explicitly manifested in the believing community. Self-surrender and not self-development is the divine order. Since man is endowed with both a carnal and a spiritual nature, he can overcome his carnal tendency only by submission to the community and with the help of his brothers.

6. All persons are born in sin and with a tendency toward evil and self-pleasure. This tendency can be modified by teaching young children the divine discipline. Children must be taught to be obedient to the "law" of their elders and superiors until they accept the mature discipline of the believing community.

7. Male dominance is reflected in the creation of the universe and is embodied in the relationships between the sexes and in marriage. Man is the leader in righteousness and in example. The divine order requires that woman be submissive and obedient to man.

8. Ultimate good is achieved in life only by surrender of the individual will to the will of God as manifest in the believing community where all material goods and spiritual gifts are shared in common. Only by living in the "ark of safety" can the individual overcome the selfish desires of his carnal tendency and be confident of eternal life after death.

Concepts Unique to the Culture

From the discussion thus far it will be observed that a number of concepts are unique to the culture. Their distinctive meanings are essential to the presentation that follows. Among the distinctive concepts are the following:

Bruderhof: The domestic group with its buildings and physical complex required to carry on the activities of communal living. It is the local biologic, economic, ceremonial, and self-sustaining unit, and is synonymous with "colony."

Community (*Gemeinsam* or *Gemeinschaft*): The practice of owning and sharing material goods in common, with each individual working according to his ability and receiving according to his need. The practice of "community of goods" (*Gütergemeinschaft*) is contrasted with individual ownership of material goods.

Carnal nature: See "spiritual nature."

Gemein: The church, a corporate spiritual and ceremonial group comprised of adult, baptized members of the community through whom the will of God is made manifest. Only men are eligible for office, and only they are represented in the formal decisions made by the church.

Leut: The word "people" used as a suffix by Hutterites to distinguish the three administrative and endogamous groups of Hutterites: *Schmiedeleut, Dariusleut,* and *Lehrerleut. Welt-Leut* or "worldy people" is also used to distinguish outsiders from Hutterites, or *Hutterleut.*

Order of God or *divine order:* The proper relationship established by God among all "created things." Man's place in the divine order determines all his relationships.

Ordnungen: The written rules of the community reflecting the will of God as made manifest through the community for particular times and situations.

Self-denial or *self-surrender:* Voluntarily renouncing individual will and accepting the will of God as made manifest in the sacred writings and as interpreted and required by the colony.

Sin: Behavioral disobedience to the will of God as revealed in the sacred writings and in the rules of community living as contextually defined by the colony. Sin is present from birth and is prone to be committed by the individual. Punishment for sin must be willingly accepted by the individual. A whole colony can sin by going against the rules and counsel of the preacher-assembly.

Spiritual nature: The unchanging, eternal, character of God in contrast to *car-*

nal nature or *human nature,* which is earthly, corporeal, self-indulgent, sinful, evil, and displeasing to God.

Temporal: The transitory and passing nature of created and material things. By contrast, the *eternal* is the never-ending and spiritual aspect of creation, including man's soul, the word of God, and God.

Word of God: The revealed will of God in the sacred writings, especially the New Testament, as supported by the moral instructions in the Old Testament and the Apocrypha, which when believed and obeyed will enable persons to receive the spiritual nature. Obedience to the word of God enables the individual to subdue the carnal nature and to conform voluntarily to the order of God in communal living.

Worldly (Weltlichkeit) or *worldly people (Welt-Leut)*: The noncolony people who live in disobedience to God and serve the earthly or carnal nature that every person possesses from birth.

2

Colony Life Patterns

EVERYONE IS OUTSIDE his apartment on the village green between the long houses and the kitchen enjoying the summer twilight. The center of the colony is alive with activity. Some of the older people are sitting on folding chairs and talking; a grandfather has a baby on his lap, a grandmother is knitting. Several young parents are bathing their preschool children in tubs set on the door step where the splashing overflow waters the flowers and washes the steps. The shop mechanic, who repaired a tractor after supper during this rush agricultural period, is scrubbing his feet in a tub by his apartment while the babysitter, his young sister-in-law, carries warm water to him. The school children dash and play leapfrog. Some young people wander around. A young boy sitting on his heels is teasing a small cluster of girls, and in retaliation one girl knocks off his hat and another pushes him off balance. There is constant movement and friendly exchange. As it grows dark the babies are put to bed, the old people go to sleep, then the school-age children and the parents of young children retire. Finally the young people return to their family apartments and the colony is quiet until sunrise.

The Hutterite view of the world is given concrete expression within the boundaries of the colony. As is common with religious men in all cultures, the Hutterites "desire to live in a pure and holy cosmos, as it was in the beginning, when it came fresh from the Creator's hand" (Eliade, 1957:65). They believe that this cosmos can be achieved only within a Hutterite colony, for only within a colony can man maintain God's order. The concept of order influences the spacial patterns of the colony; the buildings must be in proper relationship to one another and squared with the compass. Temporal patterns reflect a right order for all activity. Social patterns reflect a right order, for everyone must have the right position in a hierarchical world. There is a circular reinforcement: The world view of the Hutterites leads to the creation of an environment that is ordered spacially, temporally, and socially. It is within this created environment that the individual Hutterite matures and ages, which in turn reinforces a belief in a divinely created, orderly world. For the Hutterites, the colony, where order has become synonymous with eternity and godliness, is the center of the universe. The colony is believed to be an expression of the divine plan, and is compared to the ark of Noah in the biblical account of the flood. The people whom God selected to be in the ark were those who were worshipping him correctly. Only those in the ark (the colony) are safely prepared to escape the judgment of God and to receive eternal life. As one preacher explained, "You either are in the ark, or you are *not* in the ark."

Spatial Patterns

Dariushof is a colony located on the northern edge of the Great Plains about one hundred miles from a large trading center and ten miles from a small town which is their post office. The colony is sufficiently near to its parental colony (fifteen miles) so that work and services can be exchanged, but not near enough so that young people can walk between colonies. The colony buildings (numbering about fifty) are reached by a gravel road and cannot be seen from the public road nearby. This colony owns 6000 acres and leases 3000 additional acres, which is about the average for lands owned or leased by other colonies in the area.

Although there is no Hutterite style of architecture, there is a characteristic colony layout. The center of the colony is the kitchen complex and the long houses or living houses, with their associated sheds and the kindergarten. The long houses run due north and south, for as a preacher put it, "they are squared with the compass. You don't walk crooked to the earth, you walk straight; that is how our buildings should be, straight with the compass and not askew." Typically each long house has four apartments, each with a separate entrance. A normal sized apartment has three rooms. There is an entrance room containing a table, straight chairs, a wash basin, a cupboard for a few dishes, and the stairway entrance to the attic; off either side of the entrance room is a bedroom with two double beds, one or two day beds, and a crib. The second story of the house is one long unpartitioned attic in which the families store their out-of-season clothing and tools. Usually, there is no basement. The many other buildings in a colony are laid cut parallel or at right angles to the long houses.

All the people over six years of age meet in the kitchen three times a day to eat their meals. Here the women work preparing food for the colony, launder their family's clothes, and the colonists come to bathe or shower. Outside the kitchen is the bell that signals mealtime, announces that the women must come to work, or that calls for help should there be a fire.

Ideally the color of the buildings reflects the use of the building or the attitude of the people toward the building. In one colony the kitchen, the long houses, and the kindergarten are all painted white with blue trim. The small buildings near the long houses that are primarily for colony use, such as the bee house, the shoe shop, and the small traditional goose houses, are also painted white with blue. In contrast, those buildings used primarily for economic activity that brings in money from the outside economy are painted, in this particular colony, a bright red. They include the machine shop, the pump house, the root cellar, and various barns. An exception is the public school house, which in *Dariushof* is stucco instead of wood, is painted yellow, is oriented to face the state road rather than the colony, and from which the sign giving its former name and school district has never been removed. Although physically within the colony, the members have emotionally placed it outside.

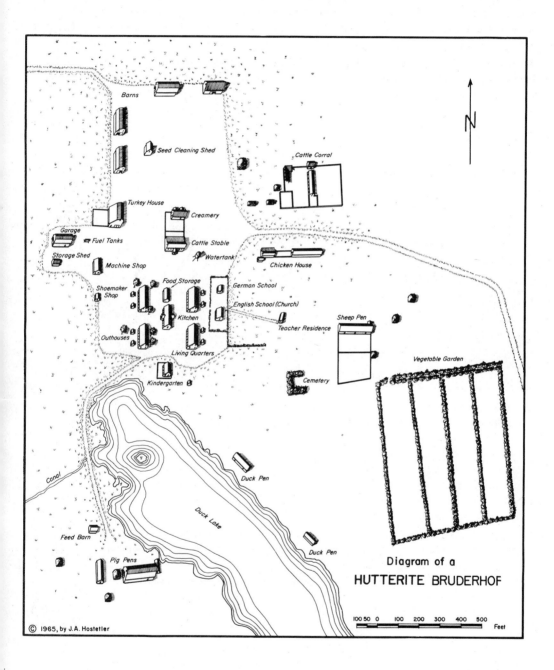

Barns

Seed Cleaning Shed

Cattle Corral

Turkey House

Creamery

Garage

Fuel Tanks

Cattle Stable

Storage Shed

Watertank

Machine Shop

Chicken House

Shoemaker Shop

Food Storage

German School

English School (Church)

Kitchen

Teacher Residence

Sheep Pen

Outhouses

Living Quarters

Cemetery

Vegetable Garden

Kindergarten

Canal

Duck Pen

Duck Lake

Feed Barn

Pig Pens

Duck Pen

Diagram of a

HUTTERITE BRUDERHOF

100 50 0   100   200   300   400   500   Feet

© 1965, by J. A. Hostetler

The spatial orientation of the buildings and the unified color scheme reflect Hutterite thinking: everything is classified; each part of the universe has its correct place, which in turn determines its correct function and proper use. "By their fruits ye shall know them" is interpreted to mean that one's appearance reflects one's attitude or the strength of one's belief. Or, more succinctly, as one strict minister said when discussing the dress of the women of his colony, "I don't care how frilly and frothy their underwear is, it's what shows that counts." What shows classifies the woman, just as it classifies the building. Her dress indicates that she is: (1) an adult woman, (2) a Christian, (3) a Hutterite, and (4) that she knows her position relative to men. It also shows whether she is dressed for work, for evening church, or for Sunday.

In the center of the colony is the communal kitchen. Like the long houses it is built on a sixteenth-century floor plan.

Hutterites consider life in this world to be transient, temporary, and of no consequence where it is lived, for they are always "in strange lands under Jews and Gentiles." In these strange lands they create their own physical environment which is remarkably uniform. The colony is the concrete expression of the Hutterite belief system and the social environment in which the beliefs are transmitted to the children. What gives a Hutterite identity is not the place he has lived, nor having lived in one or many places, but rather that in spite of geographic moves the pattern of his life has always been the same, even to the floor plan of his house and the position of his home relative to that of his neighbors. A specific place is not important —specific orientation is of utmost importance.

Communal Time Patterns

Hutterite theology teaches that there is a right order for every activity. Consequently, time functions to organize Hutterite activities and social relationships. The daily schedule involving all the members of the colony is severely patterned.

The weekly schedule builds up to a climax in the Sunday complex, preparing for Sunday. The liturgical calendar is integrated with the change of seasons and the agricultural cycle. Most women's work is organized by a time schedule, and their family work follows daily, weekly, and seasonal patterns. Their colony work is of two types: general food preparation and cleaning that is performed by the women working together in a group and the rotating colony jobs for which one or two women assume responsibility for one week at a time.

Within this small face-to-face society, time spent on earth functions as an impartial means for establishing order in social relationships. The individuals are ranked by age and sex; age determines both the group to which an individual belongs and, generally, his place within the group. The stages of life do not correspond simply to biological phases but represent social functions.

AGE SETS. Age sets are as follows:

House children: birth to two years. The small child who is still fed at home is referred to as a house child.

Kindergarteners: three to five years. The child in this age set attends the *Klein-Schul* or kindergarten. Some colonies will admit the child to kindergarten at the age of two and one-half years.

School children: six to fourteen years. These children attend the *Gross-Schul* or German school, and they also attend the public or English school from the time they are approximately six until the day before they are fifteen. At age fifteen the child enters the adult dining room.

Young people: age fifteen to baptism. This age sometimes has been called "the in-between years," or "the foolish years." Foolish does not imply any juvenile delinquency, but that the person is immature and sometimes wavering in his loyalties. Boys of this age are given the most challenging physical labor; girls do most of the colony painting.

Baptism (about age twenty) to marriage: Baptism signifies membership in the colony and is usually followed rather closely by marriage.

Marriage and adulthood: Following marriage the adult man is eligible for leadership positions, such as German teacher, chickenman, or blacksmith. After he has proved himself in these categories he becomes eligible for an executive position by being elected to the council.

Aged: The older men gradually retire from their leadership positions but remain advisers and interpreters of tradition. The older women admonish the young people, distribute the colony allotments to individual families, and help care for the babies.

Although the age sets are distinct in the life cycle, one concept may overlap several stages. For example, a child is considered a baby until he hits back, until the next one is born, until he enters kindergarten, and until he enters German school. A person becomes an adult through a series of stages: when he begins eating with the adults at fifteen, when he is baptized, when he is married, when he grows a beard, and when his first child is born. These age groups are discussed in greater detail in the chapter on socialization.

Religious man in traditional and preliterate cultures has conceived of two classes of time—the sacred and the secular—and the Hutterites participate in both of these realms. Sacred time is related to the beliefs about creation and is nonflowing, that is, it is eternal, and pertains to God who is believed to be without beginning or end. It also is characteristic of the word of God, for God's words always remain an inseparable part of him and therefore are eternal. The Hutterite soul has a beginning but no end.

Secular time is the measure of events that take place on earth, which have both a beginning and an end. It applies to all material objects, including the human body.

Intermediate between sacred and secular time are history and dreams. For the Hutterites, history is important as a dimension of the presence of God in the world. They are not interested in history in terms of dates on a secular time scale but as steps in the development of the church of God. Therefore, historical events that are unrelated to their own outlook on life are of little meaning to them and even their own history is remembered as it strengthens their faith rather than as a dated sequence of events. This means that there is some fusion of the beginning of Christianity (the historical period of Christ's birth and the writing of the Bible) with the beginning of the Hutterites (during the persecutions of the sixteenth and seventeenth centuries and the writing of the Hutterite sermons). Miracles of the sixteenth century may be mixed with those of the nineteenth, for the worldly date is unimportant compared with the fact that God "broke into history." History is a dimension of secular time that is recalled primarily because it illustrates eternity. Dreams are also intermediate between secular and sacred time. Unlike activities in secular time that should pass quickly and that one does not wish to extend, a Hutterite may report that he did not want to "leave that dream." Occasionally dreams may be equated with visions and then, too, they are outside the secular time scale.

Almost every evening, between the end of the day's work and the evening meal, secular time on the colony is interrupted by sacred time. The total community of school children and adults gathers together for the evening church service. Here they symbolically return to the time of their origin. The daily renewal of motivation through ritual participation is of utmost importance. For the Hutterite it is the sacred time which makes secular time meaningful. As the preacher slowly reads the sermons, unhurriedly recites the prayers, and the congregation sings the long, slow hymns, time is suspended: There is no rush to complete this task for the members are participating in a nonsecular time dimension. God's time is eternal, nonflowing, sacred.

In the daily round of meals and work the Hutterites are keenly aware of the passage of time. The day is broken into small units of time that form a tight, although not rigid, schedule. This severe patterning means that the individual members of the colony have little free choice and few decisions to make with regard to time. Just as material objects are not owned by a Hutterite, he also has little concept of private time. Time is not something that can be reserved for private use; it is not equated with money. The time that is needed for the completion of an operation, for example, the building of a feed shed, is considered, but not the time given by the various contributing individuals. However, the speed with which an individual works gives him status. A woman will always know how long it took her to make an article of clothing, and both adults and children set up tasks in such a way that the speed with which they are finished is obvious. There is little savoring of the moment, for the attitude toward almost everything in this earthly life is to get it over with. This applies to meals, to work, even to life itself, which is believed to be short, transitory, and of real significance only as it relates to eternal time.

DAILY SCHEDULE. *The daily schedule* involves all members of the colony, and the patterns of all individuals mesh in such a way that the colony runs smoothly and efficiently. The basic schedule varies with the time of year; it is modified in accord with the weather on a given day and is adjusted slightly to accommodate special tasks or an emergency. Thus in winter when the daylight hours are few and there is less agricultural work, the rising bell rings later (at 7:00 A.M. instead of 6:15 A.M.) and supper is earlier (at 6:15 P.M. instead of 7:00 P.M. If it rains during haying season, the rising bell may ring at 7:00 A.M. instead of 6:15 A.M. because the hay is too wet to cut. When the men are busy haying, food is sent out to the fields to them and they are fed supper when they return for the night, which may be as late as 10:00 P.M. If the women are not finished with their colony work, supper may be served late for all the adults in order that the task may be completed before the meal. When colony work is pressing, the evening church service is omitted.

Each age set has its own daily schedule. Most adults are up an hour or more before the rising bell. The time of rising is determined by their special work assignments and the age and number of their children. The woman who is baking mixes the bun dough at 3:30 A.M., the cowman starts milking at 4:30 A.M., the mother of a young baby nurses him at 5:00. At 6:15 the rising bell rings, and at 6:30 the bell for adult breakfast is sounded. At 6:45 the school children eat breakfast, and the three- to six-year-olds go to the kindergarten for their breakfast. After eating, the school girls clean the dining room and kitchen. Before the children's breakfast the women clean the adult dining room, wash the dishes, and gather food to carry back to the house children whom they feed in the apartments. The men do their chores, and the unassigned men and boys assemble informally to learn their job assignments for the morning. The work bell for the women rings any time between 7:15 and 8:50, depending on the amount of colony work. About 9:00 A.M. the adults and any children working with them pause at their places of work for a snack of a cool drink or coffee with buns. At 10:00 A.M. the bell rings to announce dinner for the kindergarten children and the house children. Dinner is brought to the kindergarten, and the three- to six-year-olds are fed by the kindergarten mother who then puts them to sleep. Mothers carry food home for their house children, feed them, and put them to bed. At 11:00 the bell rings for the school children's dinner. At 11:30 the warning bell rings to remind the men to get ready for dinner. At 11:45 the dinner bell rings, and all the adults eat their main meal, seated at their assigned places, the men on one side of the room, the women on the other, arranged around the tables according to age. Afterward the women wash the dinner dishes, the adults rest, and then have a light snack. After the snack the men return to work, and some time between 1:30 and 2:30 the work bell rings signaling the women's return to the kitchen for colony work. At 3:00 P.M. the kindergarten children are fed a snack, recite their prayers and go home about 3:30. At 5:00 the mothers carry food back to the apartments and feed their house children and kindergarten children. At 6:00 the bell rings for the school children's supper. At 6:30 the school children and all the adults assemble for the daily church service. The adults eat at 7:00. After the meal area is cleaned up and the men have finished the evening chores, the members of the colony are free to talk, visit, or work on individual projects until it is time to give the children a snack, supervise their prayers, and put them to bed. It is a schedule that keeps everyone busy, but unhurried. Time seems to pass quickly, partly because the day is divided into such small blocks of time.

The colony bell is rung to announce rising time and most of the meals; it is rung to call the women to work or to announce a wedding shivaree; its ring summons the members to help put out a fire or cope with an emergency. The bell pertains to worldly events that involve large segments of the colony. Unimportant details like snacks are not signaled by the bell, nor does the bell ring for a meal that is served directly after a church service, for the colony is already together and it would be superfluous to ring it. The bell is never used to call the members to church as is the custom among "worldly" people. If there is a question about whether there will be a church service, children are sent around to inform the families. When the minister starts to the service, the others follow. Church services are of a different order of importance from regular temporal activities; this supremely important gathering of the entire community is never announced by the bell.

WEEKLY SCHEDULE AND SUNDAY COMPLEX. The weekly pattern of activity builds to a climax in preparation for Sunday. The men are caught up in the Sunday complex only on Friday and Saturday, but the women's work and the food served follows a traditional weekly schedule. On Monday the women do the family washing at the community wash house at a time assigned to them. They wash in rotation, beginning with the oldest woman and ending with the youngest, moving up one turn each week. Ideally a woman does all the washing for her family and finishes the ironing and the mending on Monday. This is possible if she is fortunate enough to wash early that week and if there is no colony work in the afternoon. A colony may mildly boast that everyone is finished washing before 10:30 Monday morning. In some colonies the women try to polish their floors every day, but if they are unable to do this, they always polish them on Tuesdays and Fridays. During the summer the school girls pick peas for the kitchen on Tuesday morning. On Wednesday, Thursday, and Friday the women hoe the garden and pick vegetables. Women with small children wash again on Thursday. Friday is a major cleaning day. The kitchen and dining room cement floors are scrubbed with hot water and soap by all the women. The women clean their own apartments and wax and polish their floors. The cook, the baker, and the milking woman clean their respective work areas because women's work assignments change on Sunday night and they want to leave their places of work clean. One of the work teams makes noodles. Saturday morning all the women roll buns, and the unmarried girls scrub the school house so it will be ready for the Sunday church services. On Saturday afternoon everyone bathes, hair is combed, and beards are trimmed. Everything and everyone must be clean for Sunday.

There is a weekly pattern for the food. In *Dariushof* during the summer months bread is baked on Monday and Wednesday; rolls are baked on Saturday, and on Tuesday, from before seeding until after harvest when the boys are working hard, a treat is baked for the people to eat in their homes. The meals have a pattern that varies somewhat with the season and varies considerably from one colony to another. The sweetened orange pekoe tea that is served at Saturday supper is called "Saturday tea" to distinguish it from the medicinal or herbal teas that are drunk in the homes. A person with stomach trouble may regularly skip a certain meal because he knows that the food to be served does not agree with him. Sunday dinner consists of noodle soup and roast duck or, if there is no duck, chicken or goose.

Sunday begins with the Saturday evening church service. Among the *Dariusleut* the women and girls wear special Saturday afternoon dresses to this service and to supper. Saturday evening everyone sings hymns informally and visits, discussing religion and retelling Hutterite history. People try to go to bed early on Saturday night so they will not be tired for Sunday. Sunday morning the waking bell rings late at 7:30. The women wear nice dresses to Sunday breakfast, the school children wear clean clothes to breakfast, and the kindergarteners, because they do not attend church, wear regular clothes. The kindergarten children are often cared for during this time by one of the school children so that the kindergarten teacher will be free to attend church. After breakfast the women and girls change into "Bible clothes," and at 9:00 go to the service wearing dresses that are worn only for church. The sermons delivered were written during periods of persecution when the Hutterites were becoming a distinct people; they are not spontaneous or intellectual lectures on contemporary concerns. Even the rhythm of the preacher's voice differs from ordinary speech during this period of suspended temporality. The function of the sermons is not only to interpret the meaning of Biblical passages, but to guide the Hutterites through the present, secular, "evil times."

When church is over at approximately 10:30, the women and girls change into Sunday afternoon dresses and proceed immediately to Sunday dinner. After dinner the adults take a long Sunday afternoon rest. At 1:30 P.M. the unbaptized members of the colony assemble with the German teacher for Sunday school. Here they demonstrate how well they have listened to the morning sermon and recite the hymn verses they have had to memorize during the week. Families are together Sunday afternoon; there is some visiting within the colony, and if another colony is within easy traveling distance, there will often be an exchange of visits, with the visitors attending Sunday evening church (4:30) and supper (5:15) at the host colony. Weekdays are fairly uniform, but Sunday is a very different day. The leisurely Sunday schedule functions to distinguish this holy day from work days. In a Hutterite colony Sunday is long; the church service is long, the rest period is long, and there is almost no work done. Each person is reminded that God's time is measured by eternity, in strong contrast with the swift flow of hours during the busy work week.

Colonies vary in the strictness with which they observe Sunday. Some allow absolutely no unnecessary work, although none are puritanical in their definition of necessary. Other colonies are quite relaxed about observing Sunday and permit clothes to be washed, garden produce to be picked, and ducks to be plucked on Sunday afternoon. When the weather is unseasonable for harvesting during the week, a colony may decide to work in the fields on Sunday. The rule of no Sunday work that is so pervasive in many denominations in America can be adjusted by the Hutterites through a community decision made in the interests of the total colony. Since goods and time all belong to God, no individual is benefiting from Sunday work and God is still being honored.

LITURGICAL CALENDAR.   The liturgical calendar punctuates the year and divides it into definable seasons. It begins with Advent (the preparation for the coming Christ) and ends with Pentecost (the establishment of the Christian Church) [Friedman 1965:158]. During the long and uneventful winter events are placed by relating them to the liturgical calendar; during the busy agricultural season events are usually related to the major agricultural activity. Pigs are butchered "between Three Kings Day and Easter," but the baker makes Tuesday treats "from before seeding till after harvest." The liturgical cycle and the agricultural cycle are interrelated. In addition to the universal Christian holidays of Christmas and Easter there are special services that are performed at the correct time for the specific colony. The harvest sermon is read the Sunday after the harvest is gathered. Specific sermons are read to meet colony needs, such as discipline or comfort for the sick; otherwise Hutterite preachers follow a traditional yearly pattern in selecting the sermon to be read aloud on a specific Sunday. Each of the two preachers keeps a record of the date and text of all sermons delivered, who read them, and the hymns that were sung. The liturgical pattern specifies the celebration of important Christian holidays, but it is sufficiently flexible to allow incorporation of sermons that celebrate significant colony events.

Authority Patterns

All authority both inside and outside the church is believed to originate with the supernatural. Governmental authority is believed to have been ordained by God in his wrath to take vengeance on evil and to discipline the godless. Within the church there is order without physical coercion; baptized members are believed to have received the supernatural gift of the Holy Spirit through obedience and submission, and they have more power and responsibility than those who have not been baptized.

BUREAUCRATIC ORGANIZATION.   All the baptized members of the colony make up the *Gemein* or church. The group, not individuals, has the power to exclude and to accept members. Women participate in the church service by being present and by joining in the prayers and hymns and by formally greeting newly baptized members, but they "have no voice" in church and therefore do not participate in formulating colony policy nor are they eligible for church leadership positions. Only baptized men are eligible for the departmental positions such as cattleman, pigman, shop mechanic, shoemaker, and only they may vote to elect members to these positions and to decide the economic, social, and religious life of the colony.

Five to seven of the baptized men are elected to form the council. They hold the key positions in the colony, including those of first preacher, second preacher, householder, and field manager. Often the German school teacher and one or two other men who hold leadership positions in the colony or whose age entitles them to a position of authority serve on the council. The council functions as an executive committee, and makes many decisions, such as who shall be allowed to go to town or to visit relatives. The council members initiate changes in the appointment of subordinate job and departmental positions, execute discipline, and perform a judicial function. The actions of the council are directed by the church and are neither performed in the name of God nor for God, but with the help of God. Although the first preacher has the highest leadership position, his actions are subject to review by the council. Authority is thus group-centered, and decisions are derived through unified and continuous decision making by the council members. The individual council member has learned to be submissive and has been taught that the individual never reaches a state of perfection. Ideally group decisions are free from partiality and are regarded as a higher source of truth than those made by an individual.

The preacher receives no formal training prior to his election. He is elected by lot from nominations by his own colony (with the aid of other colony delegates). He is ordained to exercise full powers only after several years of proven leadership. A preacher must have the ability to be conservative in religious values but progressive in work and economic affairs. Good handwriting is an asset since he must transcribe the sermons. He must be able to exert authority wisely, since he is expected to carry out the collective will as well as God's will. His role encompasses total as well as specific responsibility. Ideally he refers all weighty matters to the council. His religious duties include conducting church services and funerals as well as performing marriages and baptisms. He hears personal problems, voluntary confession, and administers punishment for sins. He must interpret the present in terms of the past. He must oversee the life-long indoctrination of his people and the spiritual tempo of the colony. He has a direct interest in seeing that the colony schools are functioning properly, but his relation to the teachers is informal. He has the responsibility for smoothing over difficulties that arise between members. He keeps a record of births, marriages, and deaths, which reflect concern both for their history and marital patterns and keeps travel records when members leave the colony for trips. The preacher is intimately concerned with the economic well being of the colony. He keeps an eye on the activities of the householder to insure that the colony is run efficiently enough to provide for its people's needs and for future expansion. He also countersigns checks with the householder. The preacher is both guardian of traditions and spokesman for the colony in its "foreign affairs." He must remain ever vigilant against the *Weltgeist,* the spirit of the world.

The life history of the preacher of one colony reflects his intimate experience with every phase of colony life. He was born in South Dakota, the son of an assistant preacher, moved to Canada with the colony at the age of eleven. He is now in his fifth colony. In his youth he was a horseman, herding sheep and cattle, and caring for a team of horses. He drove a big steam engine for plowing wheat stubble. In the summer he swam in the James River with his peers. Some of his schooling was obtained in South Dakota and some in Canada. School days ended at age fifteen when he was assigned the job of herding sheep with his uncle. For a whole summer he slept in a tent, caring for the sheep and guarding them from wild animals, especially coyotes. Another summer he broke 6 horses, maintained the harness for all colony horses, and helped keep the plows and cultivators in repair. For five successive winters he cared for 175 horses that were fed from straw piles. If any of his charges became diseased he had to identify the animal and treat the ailment. At twenty-two he was married to the daughter of the householder; his wife was two years younger than he. Following marriage the colony was subdivided, and he and his wife moved thirty miles away with the new colony. There he held three jobs simultaneously: pigman, horseman, and tanner. At thirty-one he was elected to succeed his father as German teacher. He taught thirty-seven pupils in German school on weekdays, and fifty-six pupils on Sunday. He was elected to the council at thirty-eight and at age forty was elected a preacher to assist his father-in-law, who was head preacher at that time. By the time he was forty-three years old the colony was large enough to subdivide again. He became the head preacher of the newly formed colony that was established 400 miles away. Meanwhile there was adverse legislation directed against the Hutterites and he, with a delegation of six preachers, represented their concerns before legislative committees on many occasions. Finally, when he was forty-six his colony once again voted to form another branch, this time fifteen miles away. His leadership tasks were multiple. In one year he made ten trips to other colonies to assist in ordaining preachers, to preach funeral sermons, and to mediate disputes. Two of these were extended "trouble-shooting" trips in which the Elder of his *Leut* requested his services with six others to attend to disciplinary problems. In his lifetime he has helped to silence three preachers, two for excessive drinking and one for incompetency. He attends the preachers' conferences usually held twice annually and, in addition to his full duties in the colony, he reads Hutterite chronicles and binds sermon books during the winter months.

To summarize, the bureaucratic organization of the colony can be described briefly as follows:

1. *The colony.* The colony is a domestic group consisting typically of all persons of Hutterite parentage or persuasion residing on the premises. It is the biologic, economic, ceremonial, and self-sustaining unit within which the needs of its members are met through the activities of communal living.

2. *The church,* or *Gemein.* The church consists of all the baptized men and women. They celebrate communion as an exclusive unit and welcome back into their midst repentant members. Only baptized men may vote in formal decisions. They vote on major colony policies and determine who will hold positions of leadership.

3. *The council.* The council consists of five to seven men selected by the *Gemein* to serve an executive function. The first minister, the assistant minister, the householder, and the field manager are always on the council. Sometimes one individual will hold two of these offices or there may be no assistant minister. Except in a small colony these positions are usually represented by four different men. Generally one or two other department heads or older men also serve on the council. They sit in order of rank, facing the men of the congregation. They make practical day-to-day decisions, grant permission for travel, judge minor disagreements, and help the colony to run efficiently by making many semiroutine decisions.

4. *The informal subcouncil.* This group is so informal that it is questionable whether or not it should be listed separately. However, the first minister (sometimes

the assistant preacher), the householder, and the field manager generally meet after breakfast each day to lay out the day's work and assign men to the various jobs.

5. *The householder.* The householder is in a peculiar position for, under the direction of the council and the *Gemein* and with the help of the department managers, he is responsible for the economic prosperity of the colony. In the economic sphere he represents the colony to the outside world.

6. *The head preacher.* The head preacher is responsible for all aspects of community life, both economic and, to a greater extent, moral. He must "shepherd his flock" and "keep his hive in order." He is directed in this all inclusive role by his own colony *Gemein* and by the preachers of his *Leut* who ordained him to his position and have the power to remove him should they see fit. He represents the colony in all its aspects to the outside world and interprets the world to the colonists.

MALE AND FEMALE SUBCULTURES. Within the Hutterite colony two subcultures exist, that of the men and that of the women. Women are believed to be inferior to men intellectually and physically and to need direction, protection, guidance, and consideration. For while it is believed that man was molded in God's likeness, reflecting something of God's glory, woman was taken from man and has weakness, humility, and submission. "Women just *are* inferior." At the time of marriage the husband does not leave his colony; his bride is usually from another colony and must move into his and adjust to new people and a new life. The women do not formally participate in colony decisions nor do they select their own leaders. The man's work is assigned by the colony and is virtually unaffected by obligations to his family, except that married men do not work on the night shift during harvest and are expected to help their wives with rotating colony work if that work is heavy.

Although women do not participate in formal colony decisions as a group, they are relatively free to intervene when they or their families are affected. Women may never defend their position on the basis of being an individual, but must appeal to the welfare of the group. Such patterns of behavior can be observed with respect to the socialization of children. Hutterite youngsters who are punished by the German school teacher (always a male) seek and receive comfort from their mothers. When hurt physically, the children seek comfort from the father (the protector from the world and the environment). While the father tends to uphold the rules of the patriarchal system due to his own involvement in the structure, his wife may be more lenient in her attitude and more critical of the punishment. Wives feel less restrained in making complaints, while husbands will try to avoid confrontation with the colony power structure. A husband would rather suffer unjustly than complain openly. A wife has little to lose by complaining, but she may be frustrated because her husband will not give her the support she wants. She often projects her annoyance and mildly dislikes men as a group. Women, as a group, support each other against masculine influence, and even a dominant female will receive the support of the weaker ones and vice versa. The cultures of men and women are different and at points are opposed. Psychologists have reported striking differences in the scores of Hutterite men and women on various personality tests (Kaplan and Plaut 1958:34–44).

The loyalties of the men are somewhat divided between their peer group (the colony power structure), their family of orientation, and their family of procreation. The women are primarily oriented toward their family of procreation and therefore are more difficult for the colony to manage and integrate—especially when this attitude is coupled with some antagonism toward the men. One preacher declared: "Our colony troubles would amount to very little if it were not for the women." Married women appear to have less identification with the colony than do the men. This is indicated by the fact that men (and boys) seldom complain about the hard work even when they are haying from 7:00 in the morning until 10:00 at night. But the women complain openly when they have too much colony work in one day.

WOMEN'S WORK PATTERNS.   Women's work patterns are described in the context of colony life patterns, for the women are relatively uninvolved in economic competition with the outside society. Most of the produce they prepare for consumption is used within the colony. The men manage the money-making enterprises. For example, when ducks and geese are raised primarily for colony use, the women and children care for them; when ducks and geese (and their feathers) represent an important cash crop, a man is put in charge of the department.

Certain women have nonrotating positions of responsibility assigned to them. The most influential position is that of head cook, and the second most influential is that of gardener. The tailoress is frequently the oldest active woman in the colony, and with the help of one or two elderly women, she divides the cloth and distributes clothing and tools among the colony members, keeping a record of what has been distributed. Generally two older or unmarried women are assigned to care for the kindergarten. In some colonies there are three kindergarten mothers, and occasionally only one. The women alternate and are in charge of the kindergarten for one day at a time. In *Dariushof* the German mother is the wife of the German teacher, and the milkmaid is the wife of the cattleman. In some colonies the milkmaid also is the cook for convalescing members, especially new mothers; in other colonies the diet cook is a separate position.

With the exception of these positions, work roles for women rotate, with the change-over taking place on Sunday night. Basically the schedule is arranged according to age, but it becomes modified by childbirth, illness, visits away from the colony, and by time taken to care for a new mother. The two most responsible rotating jobs are cooking and baking. Women begin cooking and baking when they are seventeen years old and stop when they are aged forty-five or over. Other rotating jobs are helping in the milk house, cleaning up after meals, making noodles, and supervising canning. Women begin this work when they are fifteen and continue until they are forty-five or fifty. In addition to the rotating colony jobs there is general colony work that the women do as a group. This involves food preparation such as rolling buns, peeling potatoes, canning and freezing food, butchering chickens, ducks, and geese, and helping when pigs are slaughtered. The women help with food production by planting potatoes, hoeing the garden, picking vegetables, and caring for ducklings and goslings. They also pluck ducks and geese to obtain feathers that are used both by the colony to make bedding and are sold as a cash crop. The women make the soap and clean the colony buildings; they scrub the kitchen at least once a week. Girls between the ages of fifteen and marriage do most of the colony scrubbing, all the painting and varnishing, and most of the fruit picking. Young people from several colonies may be taken to distant states or provinces to pick fruit for colony use. The expeditions are considered recreation leading to wider friendship and often courtship, even though they involve considerable work.

Work patterns for women are traditional but they are modified by informal consensus of the women under the direction of the head cook. They decide when there are enough women to divide into teams for dishwashing, potato peeling, and noodle making. It is they, and not the men, who allocate certain jobs to the unmarried girls. In addition to their colony work, women must keep their apartments in

Food preparation is done by teams and on a rotating basis.

immaculate order, make the clothing used by the members of their family, do the family washing, and feed and care for their house children. One woman with eight children made over one hundred articles of clothing during November and December. The pattern of family work can most easily be illustrated by describing the work of a specific woman on a winter Wednesday when she had no rotating colony assignment. In other words, this is the basic schedule of family work to which would be added the usual Monday washing, the work announced by the ringing of the colony bell, and the rotating assignment were it a week during which the woman was responsible for cooking or baking.

Katie is in her mid-thirties. She has six children who range in age from twelve years to eight months. Her day starts at 5:00 A.M. when she takes the baby into bed with her and nurses him. At 6:15 the alarm rings and she gets up, dresses, prays, washes, and braves the subzero temperature on a trip to the outhouse. She wakes her husband who follows the same procedure, and then one of them dresses the two youngest children while the other mops the floor. The waking bell rings at 7:15 when the older children climb out of bed, dress, pray, and wash. Then the oldest daughter (age nine) rushes over to the next long house where she baby-sits for her aunt while the adults eat breakfast. Katie makes four feather beds, puts the baby on his toilet-training seat where he is watched by his eight-year-old sister, supervises the four-year-old as he dresses for kindergarten, bundles up the two year old, and just before the breakfast bell rings at 7:30, starts through the snow with her husband. The two-year-old is left with his grandfather, the preacher, who eats breakfast at home and therefore can watch the prekindergarten toddlers. Immediately after breakfast dishes, either the mother or the father gathers food to take back to feed the house children. Katie helps wash the dishes and then starts back to her house, picking up the two-year-old on the way. The moment she steps inside the door, the babysitter, her second oldest daughter, rushes off to her breakfast and the mother proceeds to feed the house children. She will not see the kindergarten child until midafternoon; she will not see the school children for more than a moment until after the adults' supper dishes are washed at about 6:30. If her husband is working nearby or working alone, he may return for a snack about 9:00; otherwise he has a snack with the men at his place of work, or in the kitchen. Katie washes the dishes from the house children's breakfast and sews until 10:00. Then she takes her pail and goes to the kitchen to get food for the house children. The two-year-old is fed as soon as she returns, the nursing baby not until 11:00. At 11:15 the two-year-old takes a nap, and the baby is nursed and put into his crib. At 11:45 the adults eat; then the women wash the dishes and return to their houses where they can rest if the house children are sleeping. After resting they have a snack or sew until their husbands come for a snack at about 2:00. Katie sews until the kindergarten children return home about 3:30. She washes them and combs the front of her hair, and at four o'clock takes her pail and goes to the kitchen to get supper for the house children and the kindergartener. She feeds them, washes their dishes, and puts the baby in his crib for a nap. At 5:00 the community gathers for church. School-age children who are babysitters are excused from church to watch the preschool and kindergarten children during church and the adults' supper. In a large colony there are two sitters assigned to each family so that each girl can take turns sitting and attending church and will not have to miss church more than half the time. In some colonies only one sitter is assigned to a family and therefore never attends church. After supper the school children must memorize their German school verses while the little children play. Between 8:00 and 9:00 the children are given a snack, washed, their prayers supervised, and they are put to bed.

During the summer Katie has a great deal more work. Cleaning is much heavier in the summer because the children and adults track in dirt and mud; in the winter shoes are removed at the door and so the floors stay clean. There is also a great deal more colony work to do. It is an unusual weekday in the summer when the women have no colony work.

Church Service and Community Integration

The day's work in a Hutterite colony is suspended toward evening when all the adults and school children follow the preacher to the school house for the evening church service. The room is large, undecorated, and almost empty except for the benches for the worshippers. Everyone knows exactly where to sit. The members of the council face the audience, with the two preachers directly behind the table. The congregation sits according to age and sex; the youngest in the front seats, the oldest on the back benches, the women sit to the right as they enter from the rear of the building, the men sit to the left. A period of quiet ushers the group into a sacred interval of time. One of the preachers draws his chair to the desk, and in a seated position announces the hymn. With his German hymnbook open, he intones the first line. The congregation repeats each line after him, singing in unison an embellished version of the tune. The voice of the minister alternates with the loud, clear, penetrating song of the congregation. After three or four verses have been sung the leader sits back and the other preacher stands behind the desk. In a quiet, slow, monotone the second man beckons the listeners to hear the word of God, and the seventeenth-century interpretation of the text is read in a stylized sermon cadence.

The sermon is centered on a passage in Rom. 8:1–6. It exhorts the reader to "walk not after the flesh, but after the spirit. For to be carnally minded is death; but to be spiritually minded is life and peace." Emphasis is on "law and order" throughout the sermon. "Since God is a God of order, He demands that order should prevail and control us in every way, so that we may be conformed to His image." The sermon states that under the rule of Moses the people of God had to obey hundreds of laws, but Christ narrowed the rules down to two: "Thou shalt love the Lord thy God, and thy neighbor as thyself. This is no law for a Christian, since he can meet this requirement without compulsion. Practicing love will do away with all enforced laws. Love works no ill to his neighbor." The sermon ends with an extraordinary descending inflection. The exhortation to prayer is given and all kneel in a forward position with faces uplifted and hands folded at shoulder height. The preacher kneels on a small stool and audibly leads a long memorized prayer which lauds the righteous acts of God the Father; he pleads for patience, endurance, and faith to withstand the trials of the earthly life. There are two more verses of the hymn and a short benediction that commends the church to God. Thirty minutes have passed since the service began. The worshippers quickly proceed to the communal kitchen.

An analysis of the Hutterite church service can function to highlight some characteristics of their life pattern. First, there is no church building. The specific place where the worship service is held is unimportant. Sacred space is not confined to one room or one building, but encompasses all the central living space in the colony; it encompasses that area where God's order is respected. Most colonies worship in the school building, some in the dining hall, others in a different room in the kitchen complex. The only requirements are that the room be large, orderly, clean and unornamented and that there be enough seats to assign everyone his place. Second, the church service contrasts with and supplements daily life. Special clothes are worn to special services, and only clean, neat clothes with long sleeves may be worn to church. Women, who seldom raise their voices when men are present, sing the hymns loudly and lustily almost drowning out the men's voices. The tempo of the service is restful and unhurried: the hymns are sung slowly, the minister reads the

Adults and school children follow the preacher to the school house for church services.

sermon in a stylized chant and the prayer is long and recited softly. There is no hurry to be finished. The community symbolically returns to the time of its origins (to Jacob Hutter, to the twelve Apostles, and to the spiritual source of its power) so that it may recreate its existence beyond secular time.

Third, the church service emphasizes the importance of right order both in seating of the members and in the sequence of the service. Fourth, the church service makes visible the authority pattern of the community and emphasizes its supernatural rightness. The women's seating is arranged by age, separate from the men. (Families do not sit together.) In the congregation men sit according to age, but on the council bench at the front of the room they sit by rank. On leaving the service the oldest man in the rear of the congregation leads. The oldest woman follows the youngest boy, and finally the council files out behind the youngest girl. The first minister leaves last, shepherding his flock. Everyone has his assigned place in church, a place that is determined by his position in the hierarchy of the colony. If anyone is absent his place is left empty and the community is incomplete. He knows that his absence is noted and that he is missed. Thus the church service corroborates the hierarchical structure of the colony, and reaffirms each individual in his God-given place in church, in the colony, and in the universe. Only he can fill his place; he is an intrinsic part of the larger whole.

Fifth, the church service not only serves a ritual function but also a didactic one. During this period the members are instructed in their discipline, their faith, their history, and the reason for their existence. Sixth, the church service is followed immediately by a community meal, by breaking bread together. Temporal bread immediately follows spiritual sustenance.

The church service reinforces the basic pattern of Hutterite life and simultaneously gives relief and depth by setting the sacred against the secular and thus permitting behavior and responses on a different plane. Within the service—protected, surrounded by, and observed by every other colony member—life becomes predictable, time is no longer fleeting, each individual is essential, and women may raise their voices in singing above those of the men. The words of the sermon are believed to flow from God and to remain part of God. Their message teaches anew and reinforces the values the full members have internalized. The church gathering is the highest moment of integration in the life of the community; it encompasses and gives meaning to all of life.

All colonies are dependent on agricultural lands and are characterized by a high degree of mechanization.

3

Technology and Economic Patterns

THE FORMAL BELIEFS about material or "created things" is that they are the gifts of God for the well being of mankind. In their proper relationship to man (communal ownership), material things "lead man to God." On the contrary, through private ownership material property is believed to "draw man away from God." The individual cannot "attain God's likeness" unless he "puts off" his carnal nature including the desire to accumulate private property (Rideman 1950:89). As in previous chapters it is important to observe the formal in relation to the informal culture pattern.

Production and Resources

All Hutterite colonies are dependent upon agricultural lands for their basic resources. *Dariushof* in Alberta was formed six years ago with a colony population of forty-nine persons. A former ranch of approximately 4000 acres was acquired by lease agreement with an option to buy. Because the ranch had been a liability to its absentee owner and a "headache" to manage with hired labor, the colony took advantage of the opportunity to secure the location and branch earlier than is normally done. Only a fraction of the land was under cultivation at that time. Annual rainfall ranges from fourteen to seventeen inches, and droughts occur occasionally. The vegetation consists of grasslands and bluffs of silver leaf bush where moisture conditions are more favorable. Soils range from black to brown to sandy loam and usually are well supplied with nitrogen and organic matter but deficient in phosphorus. With good management, rotation of crops, and use of fertilizers, arable lands produce modest-to-good grain yields. Nonarable land is good pasture. The terrain is rolling, varying from moderate hills to sloughs and water holes. The colonists say "It is a good cattle and grain country. If the loam can grow poplars and red willows, it can also produce grain." Gophers, beavers, muskrats, bobcats, lynx, coyotes, badgers, and deer interfere with the colony's productive enterprises. The gophers eat the planted grain and its roots, the coyotes disturb the sheep, and the lynx kill the poultry. Cattle are in danger of falling into the holes made by badgers and muskrats, and cropland is flooded by beavers.

Each year the colony clears from two to three hundred acres of bush land. Clearing, a seasonal activity, is an added burden for a small colony. In winter when the ground is frozen a tractor with a dozer is used to break off the small growth. The bush is heaped into windrows and left to dry for year, and then is burned. Breaking the soil is done with a large two-ton, colony-made plow. It takes three years to convert a parcel of virgin land to crop land.

The previously existing buildings on the ranch were adapted to the colony's pattern and function. The older buildings included a dairy barn, two hog barns, several grain bins, a horse barn, chicken barn, a repair shop, and two small dwelling units. The colony erected two long houses and a communal kitchen. An abandoned schoolhouse was moved into the colony site, and a kindergarten building and a seed-cleaning plant were moved in from the parent colony. Additional buildings erected included a root cellar, a smoke house, a shoe shop, a storage building with refrigeration unit, and additional buildings for poultry. With the parent colony less than fifteen miles away, it was possible for the new colony to begin with a loan of $8000, a modest amount of livestock, and use of machinery and services from the parent colony. In four years the total assets of *Dariushof* were evaluated at $360,000. The land was worth $70,000, and the balance represents the worth of the buildings and the machinery. The annual cash intake after four years was slightly more than the expenditures of approximately $74,000. While the greatest returns were realized from the sale of grain and cattle, other diversified enterprises contributing to the cash income included sheep and wool, dairy products, ducks, geese, eggs, chickens, hogs, bees, and garden produce. The colony's largest trading center is 100 miles away.

The six neighbors (twenty-four persons) are all within a six-mile radius of the colony. They are small farmers owning or operating from a quarter to a half section of land (320 acres). The colony has been willing to assist neighbors who have limited machinery and manpower to get their crops planted and harvested. County officials have generally been gratified with the positive influence of the colony. The amount of cattle thievery is believed to have declined, and the presence of the colony has resulted in added stability to the area. The colony leaders have expressed the view that the solution to the anti-Hutterite sentiment in Alberta is to locate new colonies in the north. An overwhelming sense of satisfaction is expressed by the colony members with the present location.

The small number of people and the rather atypical beginning of the colony described above contrast sharply with *Lehrerhof* formed seventeen years ago with a population of ninety persons. This Montana colony farms 16,000 acres of land. Only 4500 acres are arable, and since summer fallowing is practiced, only 2200 acres can be seeded each year. The remaining acres are grazing land and hay fields. The colony will often rent additional acreage from absentee farm owners when available. The soil is light, sandy, and shallow, and reputed to be "15–20-bushel land" in contrast to "50-bushel" lands in other nearby regions. The elevation is 3800 feet, the annual rainfall is 11 inches, and the mean annual snowfall is 33 inches. The colony is located on the eastern slope of the Rocky Mountain terrain and is within the chinook wind belt, which has a marked effect on the area's climate. Wind erosion is a serious problem, making strip farming essential.

Crops harvested in a recent year at *Lehrerhof* were 28,000 bushels of barley, 12,000 bushels of wheat, and 9000 bushels of oats. The hay crop amounted to 36,000 bales, which were stored for winter feeding. Wheat is sold for cash during the fall and winter, and all other grains are consumed by the colony livestock. The colony's acreage allotment is limited to about 700 acres of wheat, 900 acres of barley, and 600 acres of oats. If a full crop is harvested no subsidy is paid, but if the crops are hailed out or yield poorly, the colony is eligible for a subsidy. All crops, including several hundred acres of alfalfa hay, are sprayed for weeds by a colony-made, self-propelled sprayer. If allowed to grow, weeds absorb the little available moisture there is for plant life.

There are 500 head of cattle on *Lehrerhof,* and 200 are fed during the winter. A dairy herd of thirty cows provides products for colony consumption, but the cream is marketed. Five teams of horses and five saddle horses are maintained for working with range cattle and for hauling supplies during the severe winter months. The several thousand ducks that are consumed entirely by the colony are fattened during the summer. Several thousand geese, most of which are marketed in nearby towns, are raised and processed each year. Hog production is organized on a continuous basis. One barn contains the brood sows and their young, and another is used for fattening the hogs. Barns are equipped with running water, a chain drag to carry out the manure, a system of ventilation, water heaters, and a heating unit with infrared lamps to keep the young pigs warm. The colony raises about 1000 lambs each year, and sheep are a good source of income from the wool and lambs sold. Mutton is also an important source of meat for colony consumption. Sheep raising provides work for the young men on the colony during two annual periods when other work is minimal. Lambing takes place during February, March, and April, when the sheep require almost constant care.

Lehrerhof moved from a parent colony in Canada and was the ninth to locate in Montana. The present site was chosen with the aid of real-estate agents who showed colony representatives five possible locations. The advantage of the Montana site was its proximity to a good road and reasonable distance (sixty-five miles) from a large trading center. Several disadvantages of the present site have become apparent to the members. The major complaint of the colony is that the land is too poor and the rainfall too limited to produce enough crops. Several informants expressed the view that the colony should have moved to Saskatchewan, but when the group voted to go to Montana most argued: "If other colonies can succeed in Montana, we can too."

When it was formed seventeen years ago, *Lehrerhof* was considered "modern" because it acquired trucks and electricity at that time. In comparison with its own daughter colony formed after twelve years it is now "old-fashioned." "We work with outdated facilities," said the householder, "and if we had the capital we would install a modern dairy with a milking parlor, a modern poultry barn, a new hog barn, gas heating in the homes, make a larger kitchen, and buy better tractors." Since capital is lacking for such improvements and the present facilities are adequate to "get by," improvements will be modest in the years ahead.

The resource pattern of *Schmiedehof,* a third colony has many of the agricultural characteristics of the two discussed above, but is distinguished by a greater degree of mechanization. Its population numbers 124 persons. Its soil is more productive than the two colonies described, and its acreage is the least of the three colonies, or approximately 5000 acres. The productive enterprises of turkey raising, egg production, and hog production are augmented by a larger labor force as well as more modern equipment. Large chicken barns housing up to 10,000 laying hens each are automated with feeding, egg gathering, egg grading, and ventilation equipment. Mechanization appears to be more characteristic of the newly formed colonies rather than of the older colonies, provided the land is productive. Investments in machinery are allocated to the newly formed colony as a rule, and the newer types of machinery are acquired in preference to older and less efficient models. New colonies generally have a limited supply of labor and need labor-saving equipment.

The householder of a colony (who is the secretary-treasurer if the colony is incorporated) makes a yearly report to the colony members. The annual receipts of one modestly prosperous colony showed a total income of $235,000, the major sources of income coming from hogs, turkeys, geese, and eggs sold. The colony accounting showed a balance of $23,000 after expenses were paid. Should a colony have a "poor year," whether by hail, drought, or poor management, there will be very little cash available for improvements, clothing, or extras. For example, purchase of yard goods will be held at a minimum, and although new black denim must be bought for the men's work trousers, the women will be given less cloth and may not get a new dress that year.

The various enterprises are carefully considered by the colony at the yearly meeting. Whether to expand, mechanize, or diminish one enterprise, such as hog or turkey raising, will be important for the welfare of the whole colony. Important factors entering into decision making for the productive enterprises are: the cleavage between the old and young men, since the younger are more prone to support mechanization; the ability of the person in charge of a given enterprise; and the ability of the group to arrive at an amicable consensus. Since consensus is more important for "the good of the colony" than sheer efficiency, the Hutterites have refrained generally from speculative production. Their productive strategy is to maintain a wide diversity of agricultural enterprises so there will be work for everyone throughout the year. A large volume with a small but steady profit is often considered more important than enterprises yielding the largest margin of profit.

Work

The Hutterite colony is a community of work. All persons are required to work according to their ability, and the professed purpose of labor is that the needs of all may be supplied. The productive capacity of a colony varies with the size of its population, acreage, and the number of adults in the labor force. In a colony of about average population size such as *Lehrerhof,* there are 101 persons: forty-eight men and fifty-three women. The labor force (persons aged fifteen or over) consists of twenty-four males and twenty-four females; half of these 48 persons are married couples. Generally speaking, a colony with the greatest number of workers has the greatest productive capacity. All major decisions, whether they pertain to discipline or economic activity, are made by the *Gemein* and are implemented by the council. The same group makes policy for both economic and religious matters. The preacher's influence overshadows the economic life and activity of the colony. He is always the president of the colony corporation. The second preacher serves as his assistant.

Normally on Sunday evening after supper, the householder comes to the preacher's house with a summary of the work that needs to be done during the coming week. Together they formulate a plan and make assignments. Specific orders are given daily for jobs that are not routine. Two of the unmarried men are assigned to each of the large diesel tractors used for major field work. They are given the job title of "engineers" and work in shifts during the height of the field work. The older of the two persons on each tractor is "boss" and is responsible for keeping the machine serviced, overhauled, and repaired. A tractor operator has a close relationship with the shop mechanic. The combines are assigned to married men. During harvest or haying, work loads are rearranged for maximum efficiency. The cattle man, the German teacher, and the shop mechanic join the harvesting crew. Older persons, including the preacher, drive trucks or help wherever work is needed. If a tractor or combine breaks down, the shop mechanic aids the operator to get it repaired. If the two cannot repair it, the operator informs the householder who may refer the problem to the council if major costs or a trip to town are necessary. However, during rush seasons the normal steps of acquiring approval in the case of a major breakdown need not apply. A mechanic may initiate expenditures for repairs in the interests of efficiency.

The amount and type of work reveals a dynamic pattern integrated with the seasons. The leadership is on guard to provide enough work for all males during the entire year. Lack of work could mean the breakdown of a smooth running colony. In summer there is a greater need for labor then in winter. During the winter months the colony absorbs the additional labor by reassigning jobs and by giving each department additional younger apprentices. Most males have two jobs in which they have achieved some specialization. The gardner becomes a carpenter in winter. The preacher who looks after the geese in summer concentrates on book binding in winter. The pigman has an extra boy assigned to him. The carpenter has two extra helpers in winter making household furniture, and the shop mechanic has additional help for overhauling machinery.

The constant need for improving and expanding facilities absorbs much of the off-season labor. A newly formed colony with a small kitchen must build a larger complex as its population expands. Additional buildings must also be planned. During slack work periods the colony may remodel or build. The task begins with the search for a supply of lumber. Often obsolete buildings are secured from a neighbor. The building is dismantled carefully by colony labor and is transported. Bulldozing underbrush when the ground is frozen is a part of the work activity of one colony in winter. In a Manitoba colony, chasing foxes with a snowmobile and hounds is part of the winter work schedule. "Unless we get about twenty foxes each winter we will have trouble with the geese in summer," said the preacher.

In work and productive activity, competition and experimentation are desired and encouraged. Making the soil more productive, even to the extent of maintaining experimental plots, and improving farm machines are encouraged. The competition between department managers proves to be healthy for the welfare of the colony. It does not involve direct competition between individuals or their assigned status but between income-producing phases of the operations. The foreman of the chickens may accumulate a great deal of knowledge about the best methods of

Lambs are loaded on the colony truck and taken to market. Production enterprises are varied so there is work for everyone throughout the year.

building, ventilating, feeding, processing, and the exact cost of producing a case of eggs. The dairyman learns all he can from farm magazines, agricultural experiment-station publications, neighbors, and salesmen about livestock and the prevention of disease. Selective borrowing from the outside is common, and inventions in one colony will diffuse to others, often in a very short time.

The work tempo varies also with the seasons. In winter the pressure to complete any particular job is not as great as in summer, when in many colonies the growing season is relatively short. Work is not done in an erratic manner; the work pattern is predictable, regular, and satisfying. There is little effort to make work last. In reality there is often some leisure time and a minimum of physical exertion. Many work assignments are completed before the end of the day. When this happens men tend to gather in the central shop. By gravitating to unfinished tasks before the end of the day, the individual is in effect showing readiness to assist in unassigned tasks. Boys will run errands and older men will participate on a voluntary basis. The equalization of work is achieved by assignment and modified by the need for social participation after the work is done in tasks that are unstructured.

One fall day the colony decided to build a much-needed goose corral to protect the fowl from coyotes. The informal discussions about the procedures for building were very noisy, but there was no ill will or friction between disagreeing persons. After the major decisions were made, the householder said of the preacher, who was also the oldest person in the colony: "Pete knows best—he's the oldest and knows the most."

Boards were utilized from a distant pile of lumber in a preferred order. The person nailing the boards took them from a nearby pile and not from the person who carried them to the scene. Respect for the older persons, as evidenced by refusal to hurry them, slowed down efficiency. Males of all ages were present, and even three-year-old children were imitating the older ones. Six-year-old boys hammered nails into the siding with hammers. Their obstruction of the work of older persons was not resented; they were not told to get out of the way, nor were they told how to use the tools. During the afternoon several of the older men were called away from the scene to the colony's shop, and the equipment was not put away. That night it rained, and the next morning the householder was very unhappy about the tools that had rusted due to the precipitation.

Persons are very conscious of rank. Any attempt to change a person's status is met with rebuke from one's peers. The result is a cautious, humble attitude in group situations. A person who suggests a course of action must be prepared to retract his statements. Opinions are often presented passively as if they were those of another person. One obvious advantage is that a great many proposals can be considered, and the one that appeals to the majority will be adopted. The disadvantage of strong authority patterns are certain kinds of functional inefficiencies. One tractor operator, for example, did not repair his machine until it broke down because he had no authority to start repairing until it did actually break down. A younger person will not suggest to an older person that his movements are too slow, or that his talking with a stranger is slowing down the work. A young man may be afraid to propose solutions for fear of usurping authority from his older superior. A child will be punished for disobeying orders but not for breaking a dish.

At a council meeting concerned with buying a tractor, an individual who likes a particular model will present his case without appealing to personal references. He may present general ideas or the experiences of neighbors and equipment dealers. All the members may modify the proposal. No individual may claim sole credit for an idea; the group as a whole assumes authorship. This process allows all persons to share in decision making and to identify with the end product. Even though the individual does not engage in personal solitary decision making, he enjoys situations that involve group decision making.

Expansion and Branching

All Hutterite colonies are basically alike in their social organization and expansion characteristics. Each colony forms its own new colony by a planned method of splitting. A parent colony deliberately plans a new daughter colony in advance. This process of "branching out," as Hutterites call it, is made necessary by population growth and requires delicate management of capital assets; investments; redistribution of colony authority, family, and kinship factors; and work patterns. When a colony reaches a maximal population size (130–150 persons), Hutterites begin to sense management problems—problems of affluence, of inefficiency, and of supervision. The growth span between branchings varies, but the average is about fourteen years. During this time a colony may increase from about 70 to 130 persons.

The growth span of a colony is marked by successive stages from the time it is formed until it forms a daughter colony. The new colony must work hard to pay off 50 percent of its debts; the parent colony is usually responsible for the other half of the debts. A debt-free colony will expand its land holdings and obtain more equipment to provide more jobs for its growing population. A third stage is the period of affluence when a colony is able to save money for expansion, to loan money to other needy colonies, and to make and install many labor-saving devices. Each of these stages has distinct capital, management, and labor problems.

The total amount of cash required before branching, and a rule of thumb among the *Lehrerleut* is $200,000. This amount is needed for initial investment in land and equipment. To acquire this amount of capital a colony must average $10,000 of savings annually for twenty years, but some colonies will branch out after twelve years. The decision to branch out is not only an economic proposition but is dependent upon the ease of obtaining land, the population pressure, and how well the "politics" of the colony are managed. Most colonies will form a legal corporation in keeping with the laws of the state or province in which they are located. Some will incorporate when achieving a debt-free status; others will form a separate corporation at the time of branching and agree to share equally the debts for the new colony. As a minimum requirement, a colony needs about sixty persons for branching. By contrasting the new, small colony with the large colony, the different needs and solutions to communal living can be compared. Each new colony is formed in such manner that comparable age and sex distributions of the population are maintained. In the small colony of approximately sixty persons, about half are under the age of fifteen. Of the remaining adults, about fifteen are males who carry on the major work and managerial responsibility. The minimum positions in a new colony include preacher (who may temporarily serve also as householder), field manager, cattle man, school teacher, and council member at large. Five of the fifteen men are normally twenty-five or older, and only they are eligible to occupy foreman positions since they are married. The remaining ten men may be unmarried and most of them are unbaptized; their work is supervised. A number of positions are combined and held by a single foreman, until younger men become eligible. Thus the colony is characterized by much work, many potential jobs and status positions, and a minimum of struggle for the more important roles.

Both men and women tend to be overworked in a new colony. The shop mechanic has more work than he can do himself. He needs younger men who show aptitude for a number of skills like electrical work and welding. Depending on the size of the cattle herd, two men may be necessary. During the seeding and harvest seasons, great demands are made on the entire colony. At least four or five combines are necessary to harvest the crop in the short time before winter sets in. The combines are operated, when weather permits, on a twenty-four-hour basis. Several truck drivers are required to haul the grain. Others are required to swathe, and the mechanic is often busy with emergency repair work. Fuel is transported to the fields, and in case of a breakdown a mobile welding unit is brought into the fields.

New colonies seldom start with the maximum number of acres desired, or the maximum number of 6400 acres set by law in Alberta. New colonies acquire from 3000–4000 acres at the outset and increase the acreage as opportunity and need arises. Legal restrictions on buying land, particularly in a bulging colony where there is competition for roles and a need to expand the work program, intensify the internal problems. New land is secured, preferably land adjoining the colony property, but other nearby lands are also purchased as the colony increases to its maximum size. New colonies are formed within a few miles of existing colonies or as far away as 500 miles. About thirty miles is considered to be a good distance.

After a colony has reached maturity, the ratio of qualified persons to available positions is reversed. There are about thirty-five males over the age of fifteen in the mature colony. About sixteen men who are over the age of thirty hold all the important roles and status positions. Another five persons are between the ages of twenty-five and thirty and expect to be assigned to important positions soon. But since all appointments tend to be permanent, the opportunities for younger married men are limited. Thus the upward mobility that normally comes with age is thwarted by the scarcity of positions. With a larger labor force the division of labor has become more specific, overlapping jobs are eliminated, and competitive activity tends to develop along family lines. The fathers who hold important positions of power have a slight advantage in getting their sons into the favored jobs. Thus, competition reaches a stage that is no longer conducive to economic productivity but to social unrest. Work rules are more difficult to enforce where the division of labor is refined and very specific. In a young colony there is a decisive age difference between the leaders and the followers, while in a mature colony the age differences tend to disappear among people who are eligible for the important jobs and the rules are harder to maintain. Factionalization and formation of cliques harbinger the need for branching. In branching, disruptive tendencies are mitigated, and competition is channeled into expansion.

Following the decision to branch, several years may be required to form the new colony. A suitable site must be agreed upon by the whole colony, often with very limited opportunities available to them. A preferred site has the following features: availability of a good water supply and, if possible, nearness to a lake or creek; good soil and, if possible, a history of productivity; rainfall; adequate growing season; drainage; and a distinct distance from towns but access to roads and to markets. Potential fire hazards and hailstorms also must be considered. Dwellings are preferably on an elevation higher than the rest of the grounds and, insofar as

it is possible, near the center of the acreage. An equitable amount of space between living houses, the communal kitchen, the school house, and kindergarten are
important to the women. The women are permitted to view the prospective new location and to express preferences. After the land is secured, it may be farmed and
dwellings may be built over an intervening period. As a colony grows more buildings are added to an over-all layout that can accommodate additional buildings. Frequently the colony buildings are also hedged in with trees and shrubs that are
needed for protection against the wind and unwanted annoyances. Although
branching practices differ slightly among the three *Leut,* the ideals of impartiality
are the same. The following branching procedures in a *Lehrerleut* colony illustrates
the process.

Moving day had been set for Thursday, and many trucks from the nearest colonies (as far as 300 miles) came to help. The day before moving day, all members
packed their belongings. Since no one knew who was going to move, all had to be
prepared, even though half of the families would have to remain. The cattle were
divided into two herds and a lot was drawn to determine which would be moved.
The lot system was also used to divide certain kinds of small equipment. It had
been agreed that the new colony would purchase additional new equipment and
new tractors. The mother colony provided the capital for branching, and the two
colonies decided to remain a single corporation until the debts were liquidated. By
precedent the "mother" colony was to retain the older machinery. The milking
equipment was not to be removed and the new colony would obtain its own new
equipment as capital assets permitted. Benches and desks were removed from the
school, since the group staying did not have need for all of this equipment.

The preacher, with the aid of his assistant, listed families on the blackboard
in two groups, each of which was headed by the name of one of the preachers.
Family heads are permitted to choose between the two preachers. Older people were
grouped according to their known preferences, but a balance between the two
groups was worked out that took into account age, sex, family size, and relatedness.
On Wednesday afternoon, all the baptized men gathered in the school to consider
the plan of their leaders. The women anxiously waited at home to hear the out-
come. After the evening meal, the lot was drawn between the two preachers to see
which of the two groups would move. The older preacher drew the first slip of
paper from the hat. It indicated he was to remain. On Thursday morning the loaded
trucks rolled away with the assistant preacher and his group. Those on the colony
quietly unpacked their belongings, sad to be separated from friends and family, but
happy that they were allowed to remain in familiar surroundings.

Lehrerleut and Schmiedeleut rules required that a colony must have the consent of all other colonies within the *Leut* to form a new colony. This is not only a
formal courtesy but a check on the success and reputation of the communal system
as a whole. Successful communal living depends upon the well being of every newly
formed colony. If a new colony would run into difficulty with crop failure or suffer
financial reverses, every colony is obligated to come to its aid. Among the *Dariusleut,* a colony may branch without formal consent of the others, and their practices permit one of two factions in a colony to "volunteer" to move to the new location. But this "volunteering," some say, works against impartiality since it is usually

a kinship group that does the volunteering. If a colony should branch without the consent of others and experiences financial reverses, it has difficulty borrowing money from other colonies. Population growth is not the only "push" factor in branching. If the colony is divided into two factions, if the preachers, or the preacher and the householder do not get along too well, colonies will branch years before their numbers are considered sufficiently large. Branching tends to resolve the usual conflicts between brother sibships.

Branching is supported by strong authority patterns. Only the will of God is strong enough to separate family members, parents from older children, brothers from each other, and to break personal, sentimental ties. Individual desires must be subordinated to the will of the community, through which the will of God is always manifested. If the level of spiritual dedication is high, branching can be achieved with a minimum of problems; if low, then long-standing grievances can result. The highest ideal to be achieved in branching is impartiality, involving persons and property.

Consumption

"Need" and "equality" are two professed principles that govern the sharing of goods in the Hutterite colony. "Fellowship in holy things" requires sharing of temporal goods; "one person must not have abundance and another suffer want, but there must be equality" (Rideman 1950:88). Life in this world is a means to attain the goal of salvation after death. The physical existence of human life must be controlled in such a way that the divine pattern is reflected during life. The carnal appetite for food and drink, for example, must be controlled by the spiritual nature and kept in moderation. The carnal nature is believed to express itself in the adornment of the body, and in pleasures that cater to the "lusts of the eye," such as make-up, movies, television shows, dancing, and worldly music. Since these things must be avoided, they are not "needs" and not part of the consumption pattern.

The buying pattern of *Schmiedehof,* which is modestly prosperous and has a population of 124 persons, is shown in cash expenditures of $207,173 (see next page). The major expenditures go for underwriting the productive phases of the colony, for poultry feeds, building supplies, machinery required for mechanization, seeds, and insecticides. More was spent for livestock medicines, taxes (property and school), and seeds, than for groceries. Although this colony spends more than most colonies for groceries, only 2.5 percent of the total cash expenditures were for food. (The value of colony produce consumed is not shown here.) Only about 1 percent of all cash expenditures was for clothing. The $87 for fines, largely overtime on parking meters, expresses a conflict of colony time patterns versus worldly time patterns. (As a representative of the colony in town on colony business, the Hutterite operates according to colony time patterns, not acording to conceptions of privately owned time.)

The annual receipts over expenditures showed a balance of $23,000. This does not include capital investments. When allocated to individuals, this balance amounts to less than $200 per person or $1530 for each of the fifteen family heads.

Since individuals do not receive an income, no individual income tax returns are filed. However, the colony has been asked to maintain financial records for examination by federal tax agents on request. The colony receives no family allowances (applicable in Canada), no old age pensions, and no social security benefits.

CASH EXPENDITURES FOR ONE YEAR

Feed concentrates	$ 50,649
Lumber	28,125
Machinery	18,381
Gas and oil	18,360
Poults (young chicks and turkeys)	13,927
General repairs and hardware	9,835
Truck and tractor repairs	8,826
Seeds, fertilizer, insecticides	8,488
Taxes	7,739
Livestock medicines	6,756
Electrical supplies	5,283
Groceries and flour	5,281
Hospital	3,730
Clothing and yardgoods	2,261
Electric power	2,260
Meals, travel, and allowances	1,996
Meat supplies	1,970
Medical and eye glasses	1,939
Stove and plumbing expenses	1,851
Coal and propane gas	1,800
Bee supplies	1,298
Insurance and vehicles	1,004
Fire insurance	856
Bulls and boars	700
Shavings for poultry, disinfecting	586
Telephone	580
Stationery and postage	505
Donations	501
Pasture	449
Interest and banking expenses	470
Welding-shop expenses	366
Customs and brokers fees	264
Police tickets	87
Legal fees	50
	$207,173

By adapting machines to their regional needs and by doing their own repair work, Hutterites are able to offset some of the high costs of large-scale farming. Massive tractors and large trucks are repaired and maintained by colony repair shops. The labor force is trained by practical experience and by limited exposure to machine shops outside the colony. Occasionally men will take a short course in wiring or welding in a technical school outside the colony. The shop absorbs labor when idleness would be a danger to the social cohesion of the colony, thus keeping the system efficient at minimum cost. These shops utilize scrap iron and steel, often available from city dumps at a nominal cost.

The shop provides an inexpensive way to mechanize operations. Used equipment is bought, and skilled Hutterite mechanics improve and adapt it to colony needs. This applies to modernizing hog and poultry barns, including the installation of feeding and egg-gathering equipment. Hutterites often improve the existing brands of farm machinery or adapt them to their purposes. One mechanic constructed cabs of plexiglass for the grain combines long before farm implement companies sold them on the market. Another made an attachment to a combine that prevented the stones from entering the cylinder, thus protecting the machine from considerable damage. Cultivators adapted to dry-land farming are made by colonies in Canada. Specialized machines such as hydraulic rockpickers and tractor cabs are manufacture by colony shops. The Noble Blade, widely used by Canadian colonies, acts like a gigantic knife slicing the soil at a depth of about five inches. This method of cultivation, an alternative to disc plowing, reduces wind erosion. A capping knife for extracting honey from honeycombs was made by a colonist and was later patented and produced by an outside firm. Patents and royalties from such potential inventions have been attempted, but neither individuals or colonies have accumulated wealth from royalties.

The attitude toward money reflects a tendency toward saving rather than spending, accumulating it where possible, and sometimes in strange ways. When one of the authors took a preacher to lunch at a better-class restaurant in a city, the preacher was astonished to discover that the cost of his lunch was $1.75. The balance of the table conversation was dominated by comments on the exorbitant cost of the meal and a description of other restaurants where a cheaper meal could have been secured. The inability of the preacher to enjoy what we would regard as a bit of luxury is one example of the Hutterite attitude toward the use of money, though it is not typical of all Hutterites.

The taboos that are so strong against individual ownership do not apply to the economic gains for the colony as a corporate body. There are limited occasions when a colony will not take subsidies or refunds from the government. Some colonies refused compensation for missile sites and underground cables on their property. As a pacifist Christian group, their policy in this instance was not to interfere with the government and at the same time not accept payment for damages or for right of way. The colony's ability to acquire goods at minimal costs and to market products at maximum prices encompasses a wide scope of activity. A few examples follow.

Several Canadian colonies pool their orders for watermelons from a Texas dealer each year. Single colonies will take 3000 pounds and pay from one to one and a half cents per pound. Baby foxes were captured and reared until they were old enough to collect bounty. Large petroleum tanks discarded by the oil company were converted into grain storage bins. Damaged canned goods were bought at reduced prices from chain stores. Grain that had been damaged by fire was purchased from elevators for half price and fed to the livestock. Used electric motors and generators are secured from discarded sources and rewound and repaired by the colony mechanics. A Montana colony drove its trailer-truck to South Carolina to obtain earth-moving equipment from Army surplus stock. Colony bosses frequently receive courtesies from stock dealers and salesmen who covet the business of the colony. Such favors might include free subscriptions to farm magazines that directly benefit the colony. One mechanic earned for the colony as much as $70 per day repairing upholstery on trucks for outsiders.

Distribution

"To each according to his needs" is the professed or formal pattern. "Each was given what he needed according to the measure" (Rideman 1950:88). "Needs" are defined by the colony system, and "equality" does not mean that all have the same needs, nor that all receive the same amount and kind of food, furniture, or clothing. A significant aspect of the professed beliefs is that the carnal desired of the individual for property is considered so "strong" that the help of colony brothers is needed to live according to the divine pattern. The thorough distrust of individual motives built into the colony system is openly acknowledged: "In the fear of God we observe and watch one another . . . telling each his faults, warning, and rebuking with all diligence" (Rideman 1950:131–132). The Hutterite definition of need and equality varies from colony to colony, but the main features on the informal level are illustrated here.

Food is served in two dining rooms, one for the adults and one for the children. The only adults who do not eat in the dining hall are the two preachers who eat in the apartment of the head preacher. The precedent for this practice, Hutterites say, is that God-fearing people hold their ministers in high esteem (Zieglschmid 1943:363), and that it allows the leaders the necessary privacy for counsel and the opportunity for others to bring needs and complaints to their attention. Each family apartment also has a few dishes reserved for feeding preschool children and for serving snacks at coffee time between meals.

The manner in which clothing is distributed is more illustrative of the Hutterite concepts of need and equality than is food distribution. Clothing can be saved, modified, and exchanged. The head tailoress of each colony keeps a book which specifies the kinds and amount of clothing allocated to individuals and family heads. The colony book has been handed down through the years. Allotments and specifications are determined by the rules of each *Leut,* although many rules are obsolete or modified in some way. The tailoress keeps a record of the materials distributed to each family, and each mother usually keeps a record of what her family received. The yardage is distributed several times each year from the store room. A family is not necessarily obligated to use material for the purpose specified, and the material need not necessarily be used by the person for whom it was allocated. There can be trading of material between families and even between families in different colonies.

The following allotments to a newly married couple apply in *Lehrerhof.* A groom shall have from his colony: one bed with mattress and pillow, a table and two chairs, a closet, cupboard, stove, wall clock, sewing machine, and a set of Hutterite books. The bride's colony is to provide the following: 7 yards and 8 inches of bedspread material, 10 yards of comforter cover, material for a mattress pad 60 inches wide, six pillow cases, 12 yards of material for a feather comforter, a kettle, enamel dish, one cup and saucer, spoon, knife, fork, soup pail, scissors, and a large chest. Bedding as well as clothing is homemade from yardgoods. The pillows and feather bedding are usually made by the bride's family before the wedding day, especially if the bride is moving away from the colony. Feathers for making comforters were formerly allocated at the rate of sixteen pounds per person, but now a couple has the option to take a mattress and only twelve pounds of feathers. Eight pounds of feathers are allocated for each new baby.

Beds and bedding are distributed according to the number of children: with the first child the family gets a crib, with the second a *Schlafbank* (a settee that opens into a bed for two children), with the fourth child a second *Schlafbank,* with the sixth child a second bed, with the eighth child a third bed, with the tenth child a third *Schlafbank* (a bed may be optioned), and with the twelfth child a fourth bed.

Bedding for the first baby includes: a crib cover 1 by 1 yard, a mattress pad for the crib 41 by 24 inches, a pillow 22 by 19 inches, a flannel pad 22 by 19 inches, a larger pad 36 to 24 inches, and a crib sheet. For every baby the family receives two yards for shirts, four yards for two little dresses, two yards for aprons, two yards of flannel for underwear, and four yards for diapers. Obviously this is not enough for a year, but with the gifts from relatives the supply is ample. A mother may always draw upon her regular allotment of yardgoods for the family.

The rule book specifies to the inch how much yardage each person shall have on the basis of his age. Ten-year-old boys shall have material for a jacket annually at the rate of 3 yards and 6 inches. Eleven-year-old boys shall be allotted 3 yards and 12 inches, and men over fourteen shall receive 4 yards. When someone is overweight he shall have 9 additional inches. Clothing allotments change most dramatically when boys and girls become adults (fourteen years old in *Lehrerhof*). The rule book allows 9 yards of material for a girl's dress. It takes only about 5 yards to make a modern dress, but the traditional allotment is still passed out.

The amount and variety of yardgoods tends to become an object of bickering between the male-female subcultures. The colonies buy from salesmen who frequent the colonies. The tailoress knows how much the colony needs and will inquire about the price and quality. The householder must give consent to the cost, and the preacher must consent to the colors. After a clothing salesman visited the colony, one preacher said: "The women have so much material, they don't know what to do. They are just like children who don't know what they should have and who don't realize what it costs." His wife who was standing nearby denied the charge. Women admit that the material is stronger than years ago, and the men say they cannot possibly wear out everything in such a short time. Women do not hesitate to buy the less durable material so it will wear out quicker, thereby providing them with a new dress sooner. Some men and boys are finicky about colors and the fit of their shirts and trousers. Boys may refuse to wear shirts if they do not like the color.

Each household is assigned sleeping rooms according to the size of the family. A rule of thumb is one additional room for every six children. But should the colony have spare rooms, use is made of them. Frequently brothers and sisters sleep in the same room until they are twelve. The younger members of the family always sleep in a room with their parents. Every family has a box in a root cellar for storing vegetables and fruits and beverages allocated regularly to family heads. The amount of beer and wine allocated to each family varies with each colony. In one colony each person fourteen or over receives twelve bottles of beer quarterly. Wine is made by the colony from grapes and rhubarb, and each adult receives a quart per month. Rhubarb wine is in many instances regarded as "a poor colony's wine." Beverages served at weddings and at butchering time are above the family allotment. Twelve bottles of soft drink are allocated to school children (under fourteen) quarterly. Persons over fourteen who prefer soft drinks to beer may obtain twice the number of bottles.

The distribution pattern reinforces the social hierarchy. It is characterized by strong authority patterns and by practices that are adaptive. Equality is established on the basis of age and sex. Socially sanctioned needs, not individual wants, are satisfied. Impartiality and sharing are emphasized by the formal beliefs. Abuses of the system do exist, but what matters most to the colonists is not that there are abuses but that there is a right pattern from which exceptions can be granted. The important thing is that the pattern supports the sharing so that respect for the system is maintained.

Attitudes toward Property and Reciprocity

Personal property to an adult Hutterite is defined as: "Something given to me from the colony for me. Once it has been given, it is mine to use." The concept of property means the right to use but not to possess. Personal belongings include those items that are formally given to the individual by the colony, plus anything acquired by the individual with money from his small allowance. Belongings a person may have are kept in a chest under lock and key given to the individual. Personal belongings are passed on to the children and sometimes to other relatives before or after the individual dies. Yardgoods which have not been utilized for clothing are returned to the colony storehouse at the time of death. Ideally all personal gifts from neighbors or outsiders are reported to the householder. In practice, individuals are permitted to keep some of these gifts that they may have earned in return for work or favors. Some of these gifts benefit the colony. In one colony several men had received as a gift a pair of binoculars. They were used by the sheepherder and cattle foreman in their work. There is little sentimental value attached to old clothing, dishes, or furniture. Unless there is a utilitarian value, antiques are not kept for sentimental reasons but are generally worn out. Although cross-stitched decorative handkerchiefs were made until very recently by the girls for their boyfriends, the practice is dying out. The complaint is that it takes too long and it is easier to buy a gift in the store. Books are an exception to the general indifference to sentimental property. A book may be valued if it is old and if a relative has copied it. The sons of preachers who are elected as preachers have priority privileges for certain books, especially sermon books. Although having one's picture taken is taboo, having the photographs once the picture is taken (by outsiders) carries little negative sanction.

For children growing up in a colony there are two kinds of property: things under lock and key to which access is forbidden, and materials not locked up. The latter is accessible and available for use. The practice of using padlocks on the colony "is necessary to keep children from getting into everything." There is a widespread expectation that children will get into everything. One preacher said: "Opportunity makes thieves, and we have to keep things locked up." All the school girls at *Lehrerhof* from grade three to eight keep their chests locked "or else our brothers would take everything," they say. Each jealously carries her own key at all times to guard her little secret world containing chewing gum, perfume, coins, scarfs, picures, letters, and keepsakes. Boys in one colony are completely mistrusted with slingshots as windows are their favorite targets. Young people form specific behavior patterns toward property in a colony. When they are old enough to go outside of the colony and on trips to town they must learn a new set of attitudes and acquire the appropriate restraints against taking propery that is not locked up. One householder who complained about the relaxed care of tools said: "Colony people seem to think that since things don't belong to them, they do not need to be careful. They are more careless than outsiders." Mentioned were shoes, pencils, and tools. "Our colony people forget that eventually the colony must pay for everything," he said. Attitudes toward property and the colony's ability to inculcate these values vary considerably from one colony to another.

Attitudes toward property are revealed by reciprocity patterns, that is, the rules that apply in the exchange of economic and social favors. These rules of exchange also reveal the pattern of social distance, within the colony and outside of it, and the attitude of the colonists toward themselves. A scheme for understanding these relationships (Sahlins 1965:145) would include three types of reciprocity: generalized reciprocity, balanced reciprocity, and negative reciprocity.

Generalized reciprocity is purely altruistic assistance without specific obligation to return the favor. Thus, in a colony the young are nourished. The goods collectively produced are distributed on the domestic level to the individual. Individuals in the colony are the recipients of sharing and hospitality for an unspecified duration. They receive small monthly allowances for which they need not reciprocate or give an accounting to anyone. Traditionally, a small amount of spending money was given to those individuals who went to town with the colony officials to conduct their weekly business. Even though these individuals usually went to town for a specific reason like seeing the doctor, the number of people wanting to go increased. The practice was unfair to those who rarely went to town. To correct this, colonies started giving a monthly allowance to every person in the colony whether they went to town or not. The allowance varies with the rules of each colony. In *Lehrerhof,* adult men receive one dollar per month, and adult women receive one dollar every six months. In *Dariushof,* kindergarteners receive five cents, school children ten cents, and all adults receive one dollar monthly.

Balanced reciprocity is direct exchange, a two-way flow of goods and favors. Gifts or favors given require that the equivalent be given in return. Failure to return the favor means that the relationship is disrupted. In a colony, sharing is normative and to be expected. But there are additional forms of informal sharing not required by the formal patterns. These include informal friendship pacts, agreements, and reciprocal exchanges between siblings and relatives within the colony and in other colonies. Equality within the peer group is the norm, so when a whole roast duck is served on Sunday noon, the duck must be shared at the table in an equitable manner. One duck is for four persons—always the same four, since persons are always seated in the same position. Informal courtesy requires that two persons eat the breast and better pieces one Sunday, giving the other two the preferred portions the following Sunday. Courtesy also demands that the oldest person begins eating first.

Families as well as individuals may exchange goods and favors for one another. One family may exchange some of its feather allotment for a particular pattern of cloth from another family. Gift giving among friends and relatives takes many forms and is made possible by the small monthly allowances. Gifts are given to a new born baby. Candy is exchanged on many holidays. Gifts of food are prepared very quickly by a family to be given to relatives in another colony once it is learned that the colony truck or a visitor will be traveling to another colony. Greeting cards at Christmas, Valentine's Day, and Easter are exchanged by young people. Unmarried couples exchange gifts about five times each year on major holidays. Since boys have more opportunity to obtain pocket money than girls, they often give their girl friends money with which to buy gifts.

Kinship is especially relevant to modes of exchange. Close kin tend to share voluntarily and it follows that nonkin are excluded from the close circle and are regarded more as strangers. When there is a question of borrowing, brothers living together are on much closer terms than with nonrelatives. When several brothers marry into one family, there is an even greater degree of exchange. Painting an apartment is usually done in groups by the unmarried girls who are relatives, and often include those from another colony. Substitutions for the formal work pattern are usually arranged with close relatives. Persons may obtain a substitute for, but not be excused from, their regular work schedule if they go to town, attend a funeral, or suffer a minor illness. Substituting jobs among the women in *Lehrerhof* is arranged in advance and must be repaid in kind. A specific job cannot be paid back with another job. For childbirth, certain serious illnesses, or an operation, a woman is exempt from having to make up work lost. Whether help is returned depends on the closeness of kin and the amount of work performed.

Negative reciprocity, the most impersonal sort of exchange, is the attempt to get something for nothing, often with pressure or guile and is directed toward net utilitarian advantage. Among peasant societies it is often the normal thing to use strange means of seizure, including theft, haggling, and witchcraft when trading with foreign peoples. Among Navaho people, for example, Kluckhohn (1949) reports that morality is contextual rather than absolute, so that lying is not always and everywhere wrong. The rules vary with the circumstances. Among Hutterities, any tendency toward negative reciprocity is tempered by their Christian beliefs. Yet, Hutterites have been resented by shopkeepers and merchants for their sharp bargaining and parsimonious attitude. Here again, colonies vary a great deal in their practices but a few illustrations will demonstrate the pattern.

In their world view and in their history the Hutterites have held a negative attitude toward merchandizing and trading. Trading is specifically forbidden: "We allow none of our members to do the work of a trader or merchant, since this is a sinful business . . . (Rideman 1950:126). Goods may be improved through human labor, as had been done in a variety of crafts by the colonists in Moravia, but "to buy and sell again, as merchants and traders do, is sin." It is wrong in the Hutterite view because it creates profit by the factor of scarcity and defrauds those in need. The product is made more expensive and "the poor man is made a bondman of the rich." The inability of the Hutterites to appreciate the role of the merchant and the services of the middleman, reminiscent of the peasant classes in the sixteenth century, is reflected in their bargaining relationships. This is not applied equally to all outsiders, but primarily to impersonal, rather than personal, relationships. To buy from wholesalers is therefore not only more reasonable in price but also more consistent with the beliefs. The householder is under strict orders to buy at the most reasonable prices, and not to squander the funds of the church. As a good steward of the colony he must be frugal, but his attitudes may also reflect a certain suspicion of the sinfulness of merchandizing. Some of the more discerning merchants have taken the pains to observe differences between colonies in their buying practices instead of criticizing them indiscriminately. Some colonies have responded to the sensitive attitudes of outsiders and do make it a policy to buy from local merchants who can assure them quantities at fair prices. Gift giving serves as a form of compensation for minor deprivations, such as children or a young person pilfering from a neighbor. Outsiders who are friends of the colony, or persons in a position to influence the welfare of the colony, frequently receive gifts of colony produce. Extortion by outsiders also occurs, and a colony may suffer financial loss through unfair dealings. Although colonies rarely invest their money in sources other than savings banks and loans to other colonies, one colony lost one-half million dollars through an unsound venture in mining stocks.

Summary

Although social patterns are believed to be divinely prescribed and must remain stable, economic and technological pursuits vary greatly in time and from colony to colony. The production patterns consist of rational and modern methods of large scale agriculture, frequently on the more arid regions of the North American plains. A diversity of small grain, cattle, dairy, swine, and poultry enterprises are maintained by all colonies to provide cash income and to insure meat and produce for consumption. There is minimal conflict between the religious ideology and the dynamics required for collective utilization of agricultural resources. The constant need for capital expansion is intensified by the high rate of natural increase of the population.

Every person capable of working is expected to perform work assigned to him. Work patterns are clearly specified within age and sex groupings and by formal authority patterns that are firmly supported by informal associations in small groups. Men are more directly engaged in the income-producing phases of colony operations, while women are assigned to family, domestic, and food preparation tasks. There is over-all efficient use of labor groups in spite of certain inefficiencies resulting from strong authority patterns. Work is considered important, not in terms of individualistic conceptions of time, money, or man hours, but as a means of communal living.

Branching (forming new colonies) is the institutionalized response to biological growth, and the need of the communal system to maintain a small, manageable, face-to-face domestic group. The new colony is usually formed from a single over-populated "mother" colony. Branching achieves a redistribution of authority, leadership roles, work patterns, economic resources, and balancing of kinship and family influences. Every colony requires dynamic adjustments to the changing needs from its beginning until it forms a new colony.

The consumption patterns of the society are predetermined by the world view forbidding private ownership of property, the rejection of personal pleasure as an end, and an attitude of frugality rather than spending. The conceptions of "need" and "equality" are culturally defined and socially sanctioned by both informal and formal consensus on an intercolony basis. Austerity and simplicity are conciously accepted by the individual as a necessary way of life in which the carnal desire must be subjugated to the spiritual nature. Although personal and family needs contrast sharply with the North American farm family in clothing, house furnishings, and appliances, the colony as a whole manifests the characteristics of modern, large-scale, capital-oriented agriculture.

The distribution patterns reflect colony conceptions of impartiality and equality in sharing. The amount and kind of goods, determined by the rules of the intercolony association, are distributed according to the age and sex of the individual and according to family size. Although the society is communal and modern in its technology and productive features, its adherence to religious authority prevents a distributive economy based on the maximization of individual wants. Hutterite society is not only communal in production, but also in its consumption and distribution phases. Food is consumed in a communal setting. Clothing and most needs of the individual are distributed through resident household units. Profits realized from the marketplace are held by the corporation for the welfare of the whole colony.

Conceptions of property are clearly shaped by communitarian values, and by the view that material things are unimportant in comparison to spiritual things. To the individual, property means the right to use but not to possess. The formal adult relation to property is one of stewardship but, on the informal level, tools and machines are frequently neglected. Children are not trusted with valued property, and the young adult must learn the appropriate restraints on property use not expected of him as a child. Patterns of reciprocity between individuals and families are made possible through small monthly allowances to individuals from the colony. Kinship is especially relevant in the exchange of gifts, favors, and work substitutions. The Hutterite view of property forbids merchandizing or profiteering from mere exchange of goods. Gift giving to outsiders in exchange for colony favors, for minor deprivations, and as a means of expressing appreciation is normative.

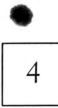

4

Socialization and Family Patterns

H UTTERITE SOCIALIZATION is a consistent, continuous process building up to two peaks in the life of the individual. Childhood is a preparation for initiation (baptism) into adult life, a period to train the child to identify with the society. Socialization during adulthood reinforces this identification and prepares the individual for death. The pattern of socialization is remarkably consistent from one colony to another, from one family to another, and from one individual to another. The system is sufficiently flexible and rewarding that an unusually high rate of success is observed. Extreme deviancy is rare. The Hutterites regard themselves as Christian believers maintaining the proper social order and not as a rationalized experiment in communal living. The continued existence of their society is secondary to obedience to God. They are, therefore, willing to become extinct as a society rather than compromise or lose the communal pattern of living, which is equated with the proper worship of God. The child is raised and the adult lives by a social pattern believed to be divinely ordained, apart from any cause and effect relationship that might be evident to an anthropologist or a worldly observer.

The function of the nuclear family is to produce new souls and to care for them until the colony takes over the greater responsibility. At all points in the socialization activity, the family supports the colony. There are virtually no limits to the number of children a couple may have except those set by nature. Having children is supported by the world view, and no birth-control practices are allowed. The prohibition of sex relations before marriage is firmly adhered to—a practice not inconsistent in a society that lives by absolutes in many aspects of life. The demographic study of Eaton and Mayer (1954), covering the period from 1874–1950, shows the median age at marriage in 1950 for women to be 22.0 and for men 23.5. Few persons remain unmarried; only 1.9 percent of the men and 5.4 percent of the women over age 30 had never married. Only one divorce and four desertions were reported since 1875. The median family size of the completed family was 10.4 children.

Intrafamily relationships are very important to the individual, for these relationships are the first ones learned by the child and will remain the primary ones throughout life. Hutterites say, *Blut ist kein Wasser* (Blood is not water), meaning that blood ties are of utmost importance and not to be taken lightly. Ego feels closer to some relatives than to others due to a variety of factors: his sex identification, similar age and interests, frequency of contact, and through other marriage ties. Ego's main relationships are with his parents, siblings, grandparents, uncles, aunts, and cousins. In line with the earlier discussion of male and female subcultures, women feel closer to the relatives of their mother, especially the mother's sisters than to the relatives of their father. Men identify more closely with the relatives of their father than with those of their mother. Boys are named after their fathers and uncles, while girls are usually named after their mothers and their aunts.

Throughout the life of the individual it is the responsibility of the whole society, not just the family, the school, the employers, or chance acquaintances to help socialize all the members. "We observe and watch over one another . . . telling each his faults, warning, and rebuking with all diligence." (Rideman 1950: 131–132). Or as one preacher explained, "It is good to live in community, for here there are always one hundred eyes watching you." The group has developed a highly institutionalized and effective system of formal education. The major levels of formal education are kindergarten (*Klein-Schul*), German school (*Gross-Schul*), Sunday school (*Suntag-Schul* or *Kinder-Schul*), baptismal instruction, and the daily evening sermons (*Gebet*). These levels of education correspond to the age sets (described in Chapter 2, page 22).

House Children: Birth to Two Years

The Hutterites are strongly opposed to any means of birth control. Man is not to interfere with the giving of life nor the taking of life—that is God's prerogative. Within marriage children are gifts from God. There is no institutional recognition of pregnancy; no special costume is worn by the pregnant woman, nor is there any modification in her work program. Informally the women help one another if one of them does not feel well, and some husbands give their wives some extra assistance when they are pregnant. The women say that the harder you work during pregnancy the healthier your baby will be. Generally, pregnancy is ignored by everyone.

In the United States where hospital costs are high, most Hutterite babies are born in the parents' apartment. The mother is assisted by a female relative and the midwife. In Canadian provinces where the Hutterites are covered by a hospitalization plan, childbirth usually takes place in a hospital. Barring actual illness or obvious complications, the mothers have little or no prenatal medical care. In contrast with the lack of emphasis on pregnancy and on the actual birth, much emphasis is placed on the postpartum woman. The relief from colony responsibilities and the full time help of a mature woman are given to the mother, not to the baby. Ideally, the woman's own mother comes from her home colony to care for her daughter. If her mother is dead or too old, an older sister is the second choice and any married sister is a third; beyond that the female relative to whom the new mother feels closest is called for assistance. During the postpartum period, the mother is cared for and mothered at the same time that she cares for and mothers her baby. The new mother does not leave her apartment. Her nurse feeds and cares for her and her house children, does all her family work and even sleeps with her at night, helping her twenty-four hours a day. During this period, the colony members and relatives from neighboring colonies come to visit the mother and to see the new baby. The visitors are given a delicious zwieback that is baked for the new mother. During the first five weeks, in addition to the zwieback and the regular colony meals, the new mother is served omelets, rich chicken soups, milk puddings and chicken roasted in butter which are prepared especially for her by the diet cook. The length of this period of complete care and dependency varies. Most mothers are anxious to return to colony participation, and as soon as their caretaker leaves they begin to eat "with the people." The integration back into the work pattern is institutionalized. Among the *Dariusleut,* on the first Monday after the baby is six weeks old, the mother starts to help wash the dishes in the kitchen; on the Monday after the baby is eight weeks old, she does the milking and if her team makes noodles she helps; when the baby is ten weeks old, she bakes for the colony and hoes in the garden if it is that season of the year. When the baby is thirteen weeks old, she cooks for the colony and is now "back in orbit" as one woman expressed it.

The care given the new mother is not only part of her continuing socialization but also affects the early socialization of the baby. During the first four weeks of life, his mother is able to devote herself exclusively to the infant's care. The Hutterites consider the neonatal child to be demanding and vulnerable, but at the same time a great joy and a pleasure to care for. The mother, who is used to being away from her children for several hours each day, finds the baby, which has in no way adjusted to the dictates of society, quite demanding. But she is confident of her ability to supply his needs and enjoys nurturing the completely dependent baby. The baby wears a cap on his head, is swaddled with a blanket wrapped over his clothes and is tied into a firm, straight bundle with a narrow cord woven just for this purpose. Swaddling is said to make the baby easier to handle and play with (that is, less easily injured). A red ribbon is tied onto the baby. Some say that this protects the baby from the evil eye; others that it keeps the baby from being fussy, from having colic, or that it just gives general protection. Even though the efficacy of the red cord is not believed by all, its use indicates the vulnerability attributed to small babies. Both swaddling and the red ribbon are usually given up by the time the mother returns to full colony participation. When the child is four weeks old, the mother's caretaker leaves, and for two more weeks the mother is relieved of colony work. Although she usually eats in the colony dining room and manages her other family responsibilities, she still has extra time to devote to the baby. The baby's schedule becomes adjusted to that of the colony, the mother leaves the child during church service and during the adult meals. As she participates more fully in the colony work program, the pattern of the child's life meshes more closely with that of the colony. The temporal pattern of colony life determines the time of feeding, playing, sleeping, and being left alone.

Nursing is accepted as a matter of fact by Hutterite mothers. No special help or modification in colony schedule is given the nursing mother other than the reduced work load for the first three months. Nursing periods are short, rarely longer than ten minutes, and pacifiers are widely used. Most babies are nursed for about a year or until the mother becomes pregnant again. Occasionally babies are not weaned until they enter kindergarten. Religious training begins with the introduction of solid food into the baby's diet (sometime between three weeks and three months). The mother folds the baby's hand in hers and prays with him before and after feeding. Religion is already a ritual, a formal activity and a social activity. It is intimately associated with food. A baby under a year will clasp his hands in the prayer position when he is hungry and sees food being carried in. At night, when the baby is put into his crib, his parents say the first children's prayer aloud to him as he is laid down.

Ich bin klein,	I am a little child
Mein Herz ist rein	My heart is pure
In Jesu Namen	In Jesus' name
Schlaf ich ein.	I go to sleep.
Die lieben Engelein,	The lovely Angels
Werden meine Wächter sein.	Will watch over me.
Amen.	Amen.

The prayer is invariably recited whether the baby is still awake or already asleep. Toilet training begins early, frequently by three months, always by the time the child can sit alone. Many children do not wear diapers after they are six months old, even though they often are not fully trained until they are in kindergarten. Prekindergarten children who are able may attend to their toilet needs outside if they show a little discretion in choosing a spot and do not soil or shed their clothes. When the child starts kindergarten he begins going to the outhouse without help.

Everyone in a Hutterite colony loves a baby. Children of both sexes will crowd around a baby to play with him and gently vie for the privilege of holding or caring for a tiny baby. A child as young as two will be rewarded by being allowed to hold a baby. When the adults are not working, the babies are always held. After supper, in a group of men sitting around informally discussing colony affairs, one or two will be seen holding babies on their laps; a grandfather may hold two or three children on his. Adult Hutterites, colony members, visitors, and everyone who passes a very young child gives him cheerful attention. The baby is spoken to, picked up, tickled, played with. However, when it is time for church or for the adult meal, the baby is unceremoniously placed in his crib and his parents walk out. Thus from the time he is seven weeks old, he alternates between being in a socially stimulating environment and being completely alone in his crib. When a child is old enough to climb out of the crib or does not generally sleep during the time his mother is away, he is either watched by a babysitter or taken to an apartment where an adult is at home. This may be a grandmother, or a crippled aunt; in some colonies the toddlers may be left with the head preacher, for he eats his meals in his apartment and therefore can keep an eye on the children while the parents have their breakfast.

A "good" baby has two major attributes. He sleeps a lot, at least during colony work periods, and he will go to anyone. In other words, he does not disrupt the colony time schedule and he accepts all colony members. A child is believed to be completely innocent until he is observed to hit back or to pick up a comb and try to comb his hair. When he hits back, or knows what a comb is for, his level of comprehension is believed to be sufficiently high that he can be disciplined. He shows both self-will and understanding. When little children quarrel over an object, the object is usually removed. If they are quarreling for another reason, they are often told to kiss one another. Some very young children will hit each other and then immediately hug and kiss one another, thus avoiding adult displeasure by quickly making up. An older house child may be strapped for refusing to go to someone other than his parents, for refusing to share food, or for being noisy and disturbing adults. He may be slapped for putting garbage in the wrong receptacle, flicked on the head or pinched for getting in the way of someone older than he, either a child or an adult. Sometimes there is an attempt, usually unsuccessful, to frighten him into good behavior. If he is teasing the geese, someone will pick up a hissing, snapping goose and chase the child with it in an effort to frighten him away. Although he is disciplined quickly and frequently, the child is considered entertaining and is petted, played with, and desired.

Hutterite mothers almost never take their house children with them when they are doing colony work. "We say that the air in the kitchen is bad for babies." Older children in the kitchen get in the women's way. During the winter the younger house children are left alone in the apartment. If there is pressing colony work at a time when the house children are not napping, the older ones may visit kindergarten or perhaps be looked after by a child who has finished kindergarten but has not yet started English school; or the father will care for the child if his colony work permits. During the summer the older toddlers, especially the boys, accompany their fathers and play near the place where the men are working, usually in the company of the youngest school boys who watch out for them and take them back to their apartments when the house children's meal bell rings. There is a united effort on the part of the colony, in which the parents actively cooperate, to wean the house child away from his parents and into the group. Most of the activities that are considered especially pleasurable, such as riding in a wagon or on the back of a truck can be enjoyed only if the toddler will leave his parents to join the fun. However, when the little ones are in a group of children, they are often the butt of much teasing.

During these first three years of life the child has passed from complete dependency on his family to a moderate degree of independence. In the summer he wanders quite freely among the long houses, around the kitchen, and visits the kindergarten. By the time the child is ready to enter kindergarten he has learned, first, that the colony takes precedence over the individual. When the bell rings, his mother leaves. Second, he has learned that the individual has little control over his environment. Punishment is usually physical, arbitrary, and inconsistent, and, from the child's point of view, often unpredictable. Third, although physical insults are unpredictable, the pattern of living is unchanging. Always the same thing is done in the same way at the same time. Fourth, he has learned to respond positively to every person (Hutterite) who comes within sight or earshot. He does not complain when he is handed from one caretaker to another. He is happy to be with people.

Kindergarten: Three to Five Years

The primary functions of the kindergarten are: (1) To help wean the child from his family and, to some extent, wean his family from him. (2) To introduce the child to his peer group, and to teach him how to function in this group. This is of great importance, for the boys will remain in this group thoughout their lives, and the girls will grow up in this same peer group. At the age when children in North American society are exerting their growing individuality and developing a concept of self, the Hutterite child is placed in a setting that minimizes treatment of him as an individual and maximizes his identity as a member of a group. (3) The kindergarten teaches the child to respect the authority of the colony in addition to that of his parents and babysitters. (4) In the kindergarten the child learns to tolerate a limited, restricted environment. (5) He is rewarded for a cooperative, docile, passive response to correction and frustration.

The first sharp change of status in the child's life takes place when he reaches kindergarten age (two and a half or three, depending on the tradition of the colony). No longer are children allowed to scream noisily when they are displeased, unhappy, or hurt. They must be quiet around adults, and even cry quietly. The kindergarten child is expected to be more obedient than a house child. A house child will sometimes get out of bed repeatedly in the evening if his parents are in the apartment and have not gone to bed themselves; this behavior is not tolerated in a kindergarten child. Visitors from other colonies do not greet kindergarten children or school children nor do they say goodby to them. Adults rarely play with kindergarten children. The status of the kindergarten child within the colony is low. He has plummeted from a relatively desirable position to the very lowest. The low status is reflected by the quality of the food served to him in the kindergarten, and by the fact that adults and older children never want him around. He is usually shooed away from any gathering, partly because he has not yet learned to be sufficiently quiet and still, but also because anyone older than he may send him away, and there always seems to be someone who exercises this prerogative. The largest number of threats are used on children of this age, for they are old enough to understand them but not yet old enough to know that the threats are empty. Outside of kindergarten the most frequently heard threats fall into two categories: those teaching that exclusion from the group is unpleasant; and those teaching that beyond the boundaries there lurks danger. Thus, a child will be told that he will be locked in the hole under the houses or given to a non-Hutterite visitor. If he opens a door the bee in the closet will sting him, the dog in the (off-limits) barn will bite him, or a bear "outside" will eat him up. When it thunders, the children are told that God is telling the children that they must be obedient.

Parents look forward to the day the child begins kindergarten. A child of this age is considered willful and useless, that is, unable to contribute any labor to the colony. "They can't do anything but memorize." The willfulness is somewhat threatening to a rigidly controlled people and because the uselessness is no longer combined with complete dependency it cannot be enjoyed by succoring caretakers. In spite of this, there sometimes appears to be an ambivalent attitude about kindergarten on the part of the mother. One kindergarten mother remarked, "We need the kindergarten, it helps the children and their mothers realize that everyone should know his or her place." In other words, it emphasizes the place of the individual within the whole colony in addition to his place in his nuclear family, and it teaches that the colony position takes precedence over the family position, that colony traditions and regulations supercede the individual wishes of family members.

The formal training of the Hutterite child before he enters English school is uniquely and completely Hutterite. The Hutterite system was evolved early in their history (during the second quarter of the sixteenth century), and has been maintained almost entirely unaffected by outside influence. In Europe during the eighteenth century, there were day nurseries where the weaned babies were cared for during the day. These children were between the ages of fifteen or eighteen months and two and one-half years. At the age of two and one-half the child entered the little school (*Klein-Schul*), which at that time was a boarding school. Since their immigration to North America, the Hutterites have not maintained boarding schools; all children sleep in the apartment of their parents, and the day nursery for the toddlers has been discontinued. *Klein-Schul* is translated here as kindergarten, which is the term the Hutterites use for it when they are speaking in English. However it should be remembered that the term as used here is not synonymous with the word "kindergarten" as used in the American educational system; it is closer to the concept of "preschool" which is used to refer to the nursery school and the kindergarten.

The kindergarten building is always an integral part of the colony. Ideally, it is near the kitchen and architecturally harmonious with it and the long houses. The small kindergarten house consists of two or three rooms surrounded by a fenced-in play yard. The house has one room that can be darkened, in which the children take their naps. They rest in assigned places, each child on his own pillow, four to a double bed or arranged along an elevated wooden platform that is built around three walls of the room. There is always a second room that functions as a dining room and, to a limited extent as a playroom. During the summer the children may be fed in a shed, which adjoins the kindergarten house. Each day the kindergarten is supervised by one of two or sometimes three older women. Who shall be chosen as kindergarten mothers is generally not difficult for traditionally they are the oldest women in the colony still able to work. Occasionally a younger woman may be assigned, but young women usually have too many house children to be able to stay away from their apartments for so many hours. The kindergarten mothers alternate, each woman taking responsibility for one full day. If one is ill she generally obtains a substitute from among her relatives. During Sunday church services, a school-age child will supervise the group in order that the adult may attend the service. The kindergarten children arrive before breakfast. They recite their morning prayer, have breakfast, recite another prayer, and sing a hymn. The children recite in unison with their hands folded either sitting on the bench or kneeling by the bench, rocking rhythmically as they recite. The teacher says the first line, and the children recite it after her. When a whole verse has been memorized the children and the teacher say it in unison. The children recite very rapidly and the older ones quite loudly. (This is the only occasion on which a kindergarten child may raise his

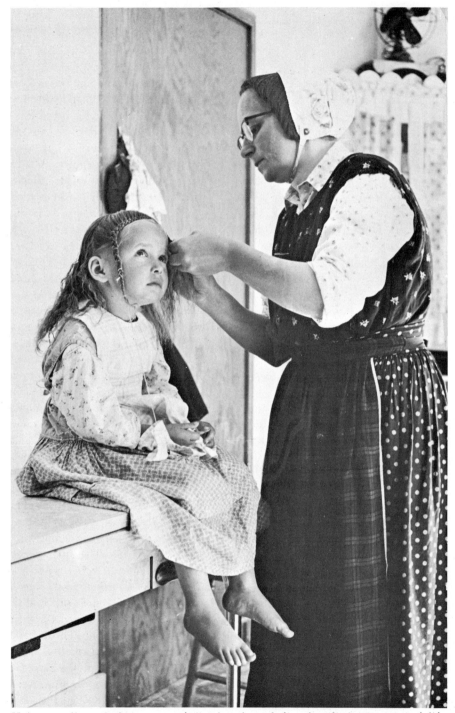

Hair as well as clothing are traditional and symbolic of a distinctive way of life.

voice.) During their years in kindergarten the children learn from ten to twenty-one prayers, depending on the colony, and from twelve to twenty-six hymns. Little if any effort is made to explain the meaning of what is being memorized.

The children are fed their dinner, the main meal of the day, at about 10:30 A.M.; then they take a long nap that enables the kindergarten mother to join the adults in the dining room for her dinner. When they awaken from their nap they are fed a snack. In most colonies they return home about midafternoon, but in some colonies they remain in kindergarten until after their supper at 5:30 or 6. The length of the school year varies among the colonies. In spite of the variation in schedule among the three *Leut,* it is obvious that one of the functions of the kindergarten is to free the mothers to do colony work.

Equipment used in the kindergartens in *Lehrerhof* and *Dariushof* is at an absolute minimum, while the *Schmiedehof* kindergarten is almost as well equipped as some small-community nursery schools. At *Schmiedehof,* children are not allowed to bring toys from home, for one of the purposes of kindergarten is to teach the children to play together, to share, and not to have personal property. "This is a time to plant unselfishness and teach the little ones to put self aside." There are no toys in either *Dariushof* or *Lehrerhof* kindergartens because "toys make the children fight." However, they may bring a pocket toy from home or a clean, light colored rag to play with. The kindergarten mother may occasionally bring a magazine or a catalog for the children. Of all the aspects of the kindergarten education the indoor play shows the greatest variation among colonies. In *Schmiedehof,* the children have an equipped playroom and, although their play is directed and somewhat restricted to the acceptable part of the room, there is considerable freedom of movement and variety of activity. There appears to be an unconscious effort on the part of the teachers to separate the sexes and encourage the acting out of future sex-determined roles. The boys tend to play at one end of the room with the blocks and tractors, the girls at the other with the dishes, table, and dolls. Little girls will play "head cook" and "mother"; the boys play "tractor," "working with animals," and "father." The teacher says there is not much trouble with the little ones fighting because she keeps them busy.

In *Dariushof,* there is no playroom and no toys are provided. When the children are kept at the table they may play with a little piece of wire, a small cardboard box, anything they might have found to bring with them. However, if these extraneous objects cause discord or get in the way of the kindergarten mother, they are thrown in the stove. Sometimes the children pick the patches off their trousers or tear out the hooks that close their jackets by having a tug-of-war between two boys who have fastened their jackets together by the hooks "to see whose mother sewed the hook on strongest." At *Lehrerhof,* the children must stay in their places at the table. Again they have no toys, but they use the snack their family brings them as a toy. Thus a piece of candy becomes a truck, they build little towers with sunflower seed hulls, or poke the hulls into an orange peel.

The outdoor play is vigorous, with a great deal of running, jumping and, when possible, climbing. This play is barely supervised; the kindergarten mother is usually inside cleaning the schoolhouse or sewing and knitting. Most of the play involves gross muscle activity. The children play horse and tractor and chase one another. Occasionally they will play child and adult and take turns pretending to strap each other. They also fight, but as long as the fighting is quiet, it goes unnoticed by the kindergarten mother. The children may be taken on walks by the kindergarten mother to visit different parts of the colony. They may be taken to see the baby pigs or to look at the new combine, or to the garden where they are allowed to pick and eat any of the vegetables or fruit. No effort is made to explain to the children what they see or to relate one thing to another. These walks are not thought of as educational experiences but as something pleasant or as a convenience for the kindergarten mother.

Asocial behavior is quickly punished; the children are not allowed to fight, quarrel, or hit; they are not to call one another names, or to use "bad words." They may not disobey the person in charge. They may not leave the kindergarten yard. If the children will not share, they are punished. The kindergarten mothers do not take the children's misbehavior personally and it does not make them tense; they assume that children so young have not yet learned how to act and they need punishment to help them learn to avoid misbehavior.

The kindergarten mothers use encouragement, praise, and rewards with the children. The little boys are told to "Eat up! eat up! so you can drive a Massey (tractor)." An older child who has been helpful may be allowed to go with the kindergarten mother to bring the food over from the kitchen. A child who has behaved very well may be the first to be dismissed in the afternoon. The group may be promised a walk if they are good, or a small child may be comforted with a little candy from the kindergarten mother's pocket. Punishment includes scolding, switching, and the use of threats. The kindergarten mothers quite often yell at the children and threaten them with the use of the willow or strap. Some kindergarten mothers feel that a leather strap is too cruel for children and makes them tough without correcting them, and that a willow switch is better. The children are given only one or two switches and those older than four rarely cry. When a three-year-old was crying after being punished, one of the five-year-olds remarked, "We're already tough, we don't cry anymore, only Michael does." (Michael had just begun kindergarten.) The children are never punished vindictively or in anger. Their kindergarten mother who, as the name implies, regards them almost as her own children, is confident that she is helping the children grow into civilized Christians.

The child's adjustment to kindergarten is eased by the fact that he has visited the kindergarten with increasing frequency since he was a toddler, that he knows all the other children intimately, and has known the kindergarten mothers all his life. In spite of this, kindergarten requires a real adjustment. The child must go to the kindergarten and he must stay there for the full day under the supervision of one adult who is not his parent. Children's initial dislike of kindergarten is considered natural and is hardly noticed. To the Hutterites it is obvious that children of this age have stubborn wills that need to be broken; and kindergarten will help teach the children not to be stubborn or willful. As one kindergarten mother explained when she swatted a child who was licking his boots, "He's only three years old and still very young. He'll need many *Britschen* before his will is broken." Sometimes a child will sit on the kindergarten bench, rocking rhythmically and repeating in a barely audible whisper, "I want my mother, I want my mother, I want my mother." The colony system is consistent and in a well-managed colony is supported by all the individuals; there is no recourse for the three-year-old but to submit. He may act frustrated for a while but finally he learns that obedience and acceptance are the best policy, and gradually he comes to tolerate his lot. He plays vigorously and with gusto, especially when he is with several other kindergarteners removed from the close supervision of any adult or school child.

The impact of the colony on the individual is greatest during his kindergarten years, for of all age groups those in the kindergarten are the most restricted, most regimented, and have the least variation in program. Physically the children spend virtually the whole day in one little building and small enclosed yard; they are cared for throughout the day by only one adult. There are no vacations during the school year except for half-day sessions on weekends and church holidays. This intensive socialization takes place at a psychologically important point in the child's development. Simultaneously, the child's parents and the adult colony members no longer tolerate the range of behavior accorded him while he was still considered to be a baby. The child can easily interpret these changes as rejection. He has fallen to the lowest status group within the colony, but he has also started the steady, rewarding climb up the steps that lead to full, responsible membership in the colony.

School Children: Six to Fourteen Years

Significant aspects of the socialization of school-age children include the following:

1. Most of the children's day is spent under close supervision by someone in a position of authority—the German teacher, a parent, or a work supervisor. During each day, however, there are periods when the child is working with his brothers or sisters and when the child is unsupervised in his peer group.

2. The groups of brothers and sisters working together within the nuclear family develop patterns of interaction that will continue into adulthood when groups of brothers will be in leadership positions in the colony and when, it is to be hoped, sisters will marry into the same colony where they can help one another.

3. Every day there are brief periods when the school children are completely free from authoritarian direction and can function as a self-contained peer group working out relationships that will last into adulthood. During the school years the children learn to function both within the sibling group and within the peer group. They learn how to adjust their dual membership as the two groups overlap, interact, and supplement one another. The two configurations persist, with traditional modification, until the peer group embraces the whole colony.

4. The school-age children are taught unquestioning obedience to Hutterite authority—their parents, teacher, the colony, to any Hutterite older than they, and to Hutterite traditions and teachings. If in the weakness of childhood they do not obey, they are taught to accept their punishment meekly.

5. They are not taught to discipline themselves, deciding what is right and following their own concept of truth; rather they are taught to do what they are told and that those in authority will watch over them, punish, and protect them.

6. The children do not develop a strong sense of guilt. Because it is natural for a child to sin and because the child is not responsible for his sin, it is not "his fault" that he misbehaves. His actions need to be directed through praise and punishment by those over him.

7. The school children master the basic ritual of Hutterite life. They abide by the rules because they are told to and they must be obedient. The children learn the verbal expression of the belief system. They must memorize the material, but it is not yet expected that they will really internalize the more difficult aspects of the beliefs.

8. The children learn to accept their proper position in the society. One German teacher begins each day's lesson with, "Dear children, content yourself gladly with a lowly place. . . . Do not interfere in things that are not your concern." The children learn to accept many frustrations passively. They learn to interpret teasing as attention, to enjoy hard physical labor, to begin to appreciate life uncluttered by material objects, and to accept the cleansing process of pain and punishment with a kind of pleasure.

During the height of the Hutterite development in Europe, the *Gross-Schul* was a boarding school and was responsible for the children twenty-four hours a day. Today all children sleep in their parents' apartments at night and a large part of the time that formerly was spent in German school is now spent attending English school. However, the children start German school before starting English school. Among the *Schmiedeleut* the children enter German school at five so they have at least one year in the German school before entering the English school. Among the *Lehrerleut* there is a subdivision of the German school called *Schulela,* which the children attend from their sixth birthday until they enter English school at about seven. There are regular class hours for the *Schulela* students that meet during English school sessions. All children are taught their German letters and read German before learning English. During the period that the children attend only German school, they eat in the children's dining room under the supervision of the German teacher, attend German school when it is in session, go to Sunday school, church, and the evening service. While the children are in English school the few younger ones help with the prekindergarten babies, or accompany their fathers around the colony. They may not go with their mothers when the women are doing colony work.

The school child can contribute some labor to the smooth functioning of the colony and can help within the nuclear family. They work as babysitters for the house children and kindergarteners, ordering their young charges around and punishing disobedience. School children are not greeted formally by Hutterite visitors nor are they introduced to adults. There are slight changes in dress when the child enters German school.

German school is taught by a married man who has been selected for his job by the *Gemein.* The position is equal in importance to any other department leadership position. Often the German teacher is on the colony council, although not invariably. It is considered an advantage for him to be on the front bench because then he can easily watch the children and note any misbehavior. The school children must sit very still and be absolutely quiet during church. If a child should fall asleep, he is not scolded, for he is still observing the ritual by being in church and being quiet. The German school mother is frequently the wife of the German school teacher (especially among the *Dariusleut*) or she may be an older woman who is appointed to the position by the council. She helps to supervise the children's meals, serves them, and teaches manners. She also teaches the girls how to clean the dining room, set the tables, wash, dry, and put away their dishes. She has no responsibility to teach any of the religious material nor does she give permission, assign jobs, and punish. She assists the German teacher and instructs the school girls in their female work roles, teaching them the pattern of work rotation that will organize their colony work for the remainder of their lives.

German school is usually taught in the English schoolhouse. Every day the children assemble in the dining room for morning prayers and breakfast. From the moment they assemble until after supper in the evening, the children are either under the direction of the German teacher or are technically released by him to do specific work such as baby sitting, helping with chores, or attending English school. During the English school year the children meet before and after English school from an hour to an hour and a half. They meet a half day on Saturday and all day during vacations or when the English teacher is sick. During this time they practice writing German script, read German, recite their memory verses from Hutterite hymns, the Book of Psalms, the New Testament, or a biblical history book. The German teacher admonishes them about their beliefs and even more about their behavior. Here rules are announced to the children, and those who have broken the rules are punished. For example, the teacher may announce that the children are not to touch the car of any English person who comes to the colony. If at a later time, any child is seen climbing on the car or playing inside it, he is reported to the German teacher who gives him the standard number of straps. This is usually three on the palm of the hand for the first offense or for a minor infringement. If a child lies about what he did he gets two more. If the offense merits greater punishment he is turned over the bench. During the long, cold winters the children wear so many layers of thick cotton flannel that when a switch is used it is more the disgrace than the physical discomfort that causes the pain.

In addition to teaching the school children to pray together, the German school teaches them to work together. Often the German teacher is the gardener or the husband of the gardener. This is a convenient arrangement, for the school children regularly help in the garden during the summer when they have no English school. The potatoes are often dug by the school children, and cucumbers and other vegetables are sorted for sale. When the children have worked especially hard or especially well, the whole group is rewarded. After they have spent the morning gathering potatoes, each child may be given a chocolate bar at dinner.

Except when they are baby sitting, all the children's work is closely supervised, for children are not expected to be able to work on their own. Each girl has an assigned task in the dining room clean-up routine for one week at a time, but some jobs are more popular than others and the children race to get these. If the girl to whom the job has been assigned does not stand up for her right or enlist the help of the German school mother to defend her place, she will have to perform one of the less popular tasks. This rushing to grab a preferred job functions to hurry the girls from the table to work, for the easiest way to defend one's job is to get there first. Jockeying for a favored position is expected among the school children, but it would be beneath the dignity of anyone old enough to eat in the adult dining room.

When the German school is in formal session the teacher holds a willow switch or a leather strap in his hand throughout most of the period. This functions more as a symbol of authority than as a ready implement for discipline; as a sceptre rather than a whip. When, for example, the preacher took over the German school and called it into session outside of regular school hours, he picked up the willow and sat down at the front of the room. The children went quickly to their assigned places and sat expectantly waiting to be told what to do. It was not merely the presence of the preacher that elicited this reaction, for he had been in the room for several minutes talking to individual children and watching them play jacks and run around the school room before he decided to take over.

The naturalness of discipline in the German school is illustrated by the following incident. Children are not allowed to speak unnecessarily during meals. (One can realize the need for this rule watching two adults serve and control fifty-six school-age children who must pray, be served, eat, and pray again—all this within fifteen minutes.) During breakfast one day two of the little girls were whispering and giggling. As the German teacher walked past he switched the noisier child with his rod. Her skirt was hanging over the edge of the bench and instead of striking the child, he hit the bench, knocking his switch under the table. The girls quickly picked it up and sat on it, hiding it under their long skirts. For the rest of the meal the teacher was without his switch. This was considered a humorous incident by the teacher as well as all the children. A single switch or being flicked on the head in the dining room, is primarily a means for directing the child's attention to his eating without having to speak to him, rather than a real punishment. The dining room is always to be quiet, except when the children are singing, and even the teacher and the German school mother do very little talking during the meal.

One of the functions of German school is to continue to teach the child to accept punishment without resistance and without anger. The German teacher uses traditional colony methods of punishment. An erring child is spoken to, is made to stand by his desk, must sit on the front bench, may have to kneel, may be sent to stand in the corner, or must come forward and receive three straps on his hand. All these punishments function primarily to remove the child from the group and shame him in front of his peers. Other punishment may be given in front of the group or after German school is dismissed, depending on the discretion of the teacher. For repeated offenses a child is first spanked with a switch and, if he persists, with a leather strap. If these methods do not work, he is placed with the youngest children in the German school until he is ashamed and claims by proper conduct his age-determined place in the hierarchy. Praise is used to encourage all the children, even the slowest.

Children's native abilities are taken into account and children who are slow-learners are given less material to memorize. German school is "ungraded" in the modern sense, with the children progressing at their own speeds. There are no grades or formal levels, rather an accepted sequence of material to be learned. First they learn to recognize their German letters both in Gothic print and Gothic script; they learn to recognize numbers. Then they learn to write the letters and the numbers. Next they are taught to recognize common syllables; these elements they combine together to form words. After several months they are reading and writing the medieval German script. Throughout their school years they practice handwriting, memorize prayers, hymns, Bible stories, the catechism, and episodes of Hutterite history. The children also learn the directions of the compass, measurement equivalents, and how to write to ten thousand. The children are taught to work efficiently; they know what to do, how to do it, and when to do it, but they are never given an opportunity to ask if it should be done at all.

Within the system there is no room for doubt. The intellectual content of the German school curriculum is but a small part of the total learning that takes place in the school. The primary function of the German school is to teach the Hutterite children the ritual of life, which applies primarily to two different areas. The first is ritual that insures the smooth social functioning of the group in all the details of everyday interaction. The second class of ritual reduces the fear of death and physical injury. The children in the German school learn primarily the first type of ritual although the second is not neglected. Psychological studies of Hutterite adolescents show they are not frightened by threats to their bodily integrity although there are in fact many instances of physical injury (Hostetler 1965:79). These are not anticipated nor do they serve as a source of worry.

One of the tasks of the German teacher is to teach his charges table manners. They learn to eat with a knife and fork in addition to a spoon, to serve themselves from the nearest quarter of the serving plate, the oldest taking the first helping. The table rules that the children memorize are very similar in form (rhymed couplets) and content to those widely circulated in Europe during the fifteenth and sixteenth centuries. These rules dictating how they are to behave at the table (*Gesang-Büchlein* 1961:86–89) are memorized and recited in unison.

Within the larger context of ritual the children learn the verbal content of their religion. They will internalize the theological and moral content when they seek baptism. At this stage in the child's development he is taught to avoid punishment at the hands of supervising adults. Children memorize the material because they must be obedient; only when they grow older are they expected to understand the concepts they recite. This pattern of learning is consistent with that observed in other cultures having an oral tradition. Wisdom, which is preeminently social rather than technological, is first memorized then understood.

School children learn a great deal about the authority structure of their society and how to live comfortably within it. They have internalized the Hutterite hierarchy that gives precedence to age, but they have also learned that the authority of each individual is limited. Therefore, the child knows when and whom he must obey and what orders can be safely ignored. If the children are disturbing the young pigs and an adult tells them to stop, they pay little attention other than to stay beyond reach of the reproving adult. If the pig boss tells the children to "get away," they leave.

The German teacher's role is defined by tradition; his work is watched over by the council members and noted by all the members of the church. Concerning his task, one German teacher wrote: "I feel that the greatest challenge in my work is to put a good religious foundation under the children so that they may become respectable and honorable members of the *Gemein*. I always like to think of them as young tender plants in the Garden of the Lord where the school teacher's duty is to trim, weed, and water as he finds it necessary. Of course I realize that neither the plant nor the waterer can achieve anything without the Lord's blessing."

The peer group which functions beyond the supervision of the German teacher and the child's nuclear family, contributes to the socialization of the individual. Every day, while the adults are eating, the children of the colony who are younger than fifteen are left with no adult supervision. During these few minutes the children are virtually free to settle their own disputes and to apply peer group pressure on deviants.

Boys and girls are taught to play differently. When a group of girls played leapfrog ignoring their long skirts, the preacher picked up a switch and walked toward the scattering group admonishing firmly with "girls don't play that way." In some colonies the sisters in each family have an assigned play house. This may be the family's goose house, which the little girls use when there are no goslings in it, or it may be an extra shed or even an infrequently used room in one of the houses. Here they spend hours playing with each other and with the girls from other families. Usually each little girl brings a doll out from hiding under her skirt and plays house happily knowing that she will be undisturbed by the preacher who would throw the doll in the trash were she to flaunt it. Occasionally the girls will play that they are "English" women, wearing funny clothes, strutting about, and talking loudly. On the very rare occasions when one of the girl's brothers is allowed into the play house, the children may play "doctor," drawing realistic incisions complete with sutures on one another's arms. The oldest sister in each family keeps the key to the play house and children are free to exclude the children with whom they do not wish to play. Boys are customarily excluded; among the girls, the play groups constantly shift and there is almost always at least one girl who temporarily is being locked out of the favored play house. Sometimes the girls divide into cliques, each clique using one play house. The cliques, so separate in their play, quickly unite to confront the boys or an adult. Most of the girls' play is near the long houses and the center of the colony. In contrast, the boys range much further in their play. The boys are more individualistic and often will work or play alone. The girls demand detailed conformity within the clique or the dissenter is excluded. Only by adjusting completely to the group may a girl remain to play.

Much of the children's play is physically vigorous and often rough. They run and chase and climb. They fight hard, quickly, and quietly. They vie with one another, showing no physical fear jumping off high places or pushing one another in front of the tractor. Adults ignore the children's dangerous play; they are busy, and the children are the German teacher's responsibility. If an adult should not like the play, he might yell at the children, who would momentarily scatter. If he were really bothered he might tell the German teacher who, if in agreement would make the children stop their play. Normally, however, no one notices how the children play if they are quiet. School children play such games as "whistle when it hurts," in other words, games that are exercises in discomfort and endurance. The free play of school children reinforces community values: the children learn to ignore physical discomfort and fear of injury and to minimize the importance of the body; the changing play groups teach the unpleasantness of being excluded.

During the school years the children identify closely with their respective peer groups. When the group works hard or some of the members do especially well, everyone in the group is rewarded. When most of the group misbehave, everyone is scolded. The child learns not only that the behavior of the group directly affects the rewards and punishment he receives, but he also learns that peer-group solidarity can protect him from "outsiders" (adults) and from punishment. Transgression can go unnoticed if no one reports it. If no one will tattle on another, often the transgressor is protected; it generally requires too great an effort to punish every child. However, the school child also learns that his own peer group can punish him even more severely than does adult authority. The rules of the colony can sometimes be circumvented, even more often the rules of the family can be ignored; but the child cannot ignore his own peers.

All assigned babysitters in the colony are school children between the ages of six and fourteen. Usually only girls are assigned to a specific family but, if needed, boys may be assigned; however, all boys baby-sit informally within their nuclear families. The primary functions of the babysitters are to keep their charges quiet and out of the adults' way and to protect them from serious injury. The sitters enjoy being able to boss their charges and to punish them, but they do not like the work involved. The babysitters are highly responsible in that while they are in charge, they never leave their charges unattended. When there is a conflict, the older child is supported against the younger. During the summer there is some exchange of babysitters among colonies. Older school girls may be sent to help a sister or an aunt who has married into another colony. These girls help with the young children and with the family work and attend German school. Boys of this age do not perform comparable exchange work: it is not considered right for a young boy to be away from his father or his home colony. For a girl it does not matter, because she will probably leave her home colony when she is married.

A child is never punished by depriving him of food. When an English teacher kept a child in school during the dinner period, the German teacher went to the school house and got the child, firmly explaining to the English teacher that children, no matter how they have behaved, need to be fed. Work is never used as a punishment, for there is no colony work that is categorized as unpleasant and everyone must be willing to do his share of any type of work to which he is assigned. Privileges are not withheld from a naughty child; if the group goes on an excursion to help butcher sheep or brand cattle, he goes too. If it is his turn to visit another colony when the truck is going, he may go no matter how recently he has misbehaved. When a child is observed to be misbehaving, he is punished immediately; if misbehavior is reported to the German teacher or to his father, the child is punished at the first opportunity. If a child is strapped by his father, he is comforted by his mother the moment it is over; if his mother is not present, his father may comfort him after strapping him. It is believed that the punishment removed the misbehavior, and the child is fully accepted without having to atone further. When a child is punished in front of school children with no parent present, he is not generally comforted, unless it is at the end of the session and he can return home. Occasionally if a parent is considered to be too lax with a child, the father may be asked by the colony to punish him at home, or to punish him in the presence of the council. There is a basis for this practice in the biblical story of Eli recounted in one of the Hutterite sermons on child training. If a father, especially one in a high-colony position, feels that his child is not being properly disciplined, he will take over the task. In one colony the English teacher was almost unable to manage one of the larger boys. She did not report this to the German teacher, but the boy's father learned of the problem. The next morning he came into the class room with the child, sat on one of the church benches during the opening exercises and then asked the teacher, "Has John been misbehaving in school?" The teacher answered, "Yes." The father went to his son's desk, turned him over it, strapped him in front of the children, and then he walked out. Punishment is considered part of the pruning that is needed to shape the individual to fit into colony life.

Children are expected to misbehave when they are unwatched, for that is believed to be their nature. These same children are expected to grow into responsible colony members. Misbehavior during childhood does not endanger the child's future reputation nor foretell an unsuccessful adulthood. Throughout his life, the expectation is that the child will become a more responsible, highly socialized person. The school-age child must accept his place in the social structure of the colony, respecting those older than he and caring for those younger. It is not his responsibility to care for himself nor to discipline himself. The same child who is completely responsible in his assigned colony job may get into serious mischief during unsupervised play with his peers. Thus, if several children play with fire, accidentally setting the barn aflame, they are punished but not made to feel guilty, for lack of judgment and misbehavior are natural to children. It is not the responsibility of the individual but of the adults and of the system to protect the children from their natural instincts.

Childhood comes to an end with the fifteenth birthday. The young Hutterite leaves the children's dining room and is said to be *bei die Leut,* with the people. The *Mandle* (little man) has become a *Buah* (boy); the *Dindla* (little woman) has become a *Die-en* (girl).

Young People: Age Fifteen to Baptism

On the eve of his fifteenth birthday the child goes, alone, to the German teacher. For weeks he has been counting the days until his birthday with eager anticipation, but the last few days the child frequently becomes apprehensive about the impending change. Those older than he tease him, saying that the German teacher will give him a whipping he will never forget. The child does not really believe the teasing, but the change in life patterns is both enticing and frightening to him. The German teacher exhorts the child to be good, to work hard, to show good manners, and to do quickly and pleasantly everything that an older person asks him to do. On leaving the German school, the individual accepts the responsibility for saying his evening and morning prayers without supervision. The person is given a catechism and several Hutterite books for his personal use.

The child takes his place in the adult dining hall, for he is now "one of the people." The boy sits with the men, occupying the "lowest" position at the table. The girl sits with the women in a similar relationship. She must wait on the men, should they rap for anything during mealtime. The young person will do assigned colony work with adults of his own sex. He no longer does the work of children such as baby-sitting, gathering potatoes, or cleaning up the colony grounds. He no longer attends English school. He is greeted with a handshake by visiting Hutterites and is allowed into the fringe of adult social life. There is no ceremonial recognition of the person's new status, for such a recognition would give undue emphasis to the individual.

The young person is in a transitional stage from childhood to adulthood. The colony recognizes both aspects of his personality. Physically he is considered to be an adult who is capable of hard work and of working with adults. Religiously he is considered to be a child who must attend Sunday school and must memorize his weekly verses. Emotionally he vacillates, and this period is sometimes called "the in-between years" or "the foolish years," meaning that the loyalties of the individual have not completely crystallized. Some disregard of colony mores is expected during this period, but moodiness or poor work performance is not tolerated. A good young person is "always obedient and never talks back."

Soon after his birthday, but gradually and at the colony's convenience, he is given various gifts that reflect his change of status and are needed in his new role. Both boys and girls are given a locked wooden chest in which to keep their personal belongings. The boy is given cloth for good suits and shirts, the girl receives material for dresses. Boys are given work tools that they are responsible for keeping in good working condition: a spade, a pitchfork, a hammer, a saw, and in some colonies a spoon with the individual's name inscribed on it. The girls receive tools that they will care for and use for colony work: a scrub pail, a paint brush, a hoe, kitchen knives, a broom, knitting needles and in some colonies a rolling pin and, until recently, a spinning wheel.

For two years the young person is in an apprentice position. He is not given responsibility for an expensive machine nor for work that, if it should go wrong, would cause a great deal of inconvenience or much money lost. A boy is usually assigned the responsibility for a tractor at about the age of seventeen or eighteen. It is his to use for the colony, to care for and keep in good condition. Formerly boys of this age were given a team of horses to train for field work. At the age of seventeen, girls begin to take their turn baking and cooking for the colony, being responsible for all the food prepared or all the baking done during the week. The boys and girls of the in-between years constitute a mobile labor force that can be used throughout the colony as needed (in jobs suitable to their sex) and they also may be sent to other colonies to help during a time of need or stress. The boys in this group supply most of the hard labor and enjoy the opportunity to demonstrate their strength and stamina.

The young person is subject to the control and influence of the colony, his family, and his peer group. The influences of these three groups are less well integrated during the years preceding baptism than at any other period. Because the areas of control are more diversified, the young person has slightly more freedom. His work is under the control of the colony, although there may be some tendency to let the son work with the father and the girl work with her mother or with a sister. While working, the boy is under the direction of the departmental director who may send him off the colony on an errand. His religious development is under the direction of the German teacher, who is in charge of Sunday school. In areas of moral and social behavior the young person is primarily the responsibility of the German teacher in some colonies, while in other colonies he is the responsibility of the preacher. Depending on the degree of infringement, the transgression may be handled by the German teacher, the preacher, or the council. Permission to leave the colony, other than to do an errand needed for work, is granted by the German teacher among the *Dariusleut* and the preacher among the *Lehrerleut*. It is not granted by the parent. The colony permits dating among the young people, although a specific family may forbid it. If a young person is dating someone of whom the colony disapproves, such as a first cousin or the English school teacher, the German teacher or the preacher speaks to him; if he is dating someone whom his family dislikes, it is a family responsibility and the colony is not involved. Both the colony and the family expect a certain amount of deviant behavior. Much of this takes place within the peer group who participate together in forbidden activities of singing English songs and playing mouth organs, but the peer group tolerates only certain approved activities. If an individual deviates too radically, he is excluded by his peers.

Within the family, the young person is no longer grouped with the school children. His parents take a greater interest in his wants and needs and identify somewhat more closely with him. The relationship is still completely hierarchical, but now that the child has become a person and the German teacher is no longer responsible for virtually all the child's waking hours, the parents concern themselves more with their child's free time. "A wise mother has always an odd job for the idle hands of her daughter." Parents have almost complete veto power over the marriage of their children, and they start early to express interest in this aspect of their children's lives. Among the *Dariusleut,* physical punishment is not used by the colony representative after a child is fifteen (for girls over twelve it is not used, other than straps on the hand), but it may be used by the parents; however, in practice this is not necessary. Among the *Lehrerleut* a disobedient daughter may be locked out of the apartment and forced to seek lodging with another family for the night. (The parents know that she will not leave the colony and that other families will give her shelter.) Mothers continue to teach their daughters homemaking skills and girls of this age often take over the family laundry and do much of the sewing.

The young people in the family support one another. The girls may sew for their brothers, making clothes for them that deviate slightly from the accepted pattern. An adult woman would not make a shirt with a forbidden pocket on it, but a young girl will make such a shirt for her brother (or boy friend) or will add a pocket to a shirt he already owns. A sister will iron her brother's clothes just the way he likes them. Boys have some opportunities to earn extra money that they may lend or give to their sisters or girl friend. Brothers in one family frequently will date sisters in another family. Brothers support one another, and sisters support one another during the period of dating and after they are married. It is common to find marriages where several brothers are married to women who are sisters. The cooperative patterns learned in the nuclear family influence the pattern of courtship. Colonies that frequently visit back and forth and exchange work and produce and from whom marriage partners are chosen are usually ones that are closely related by family ties.

Although the young person has left German school and English school, his education is continuing. The family continues to teach its sons and especially its daughters skills helpful for colony life. In the religious sphere the formal education of the individual continues, for the young people must attend Sunday school on Sunday afternoon. Sunday school functions to reinforce the sermon and to discipline the young people to listen to the sermon. Although all the German-school children also attend and participate in most of the program, the emphasis of the Sunday-school program is directed to the young people who have finished the colony German school but who have not yet been baptized. The colony punishes young people for moral or social transgression by admonishing them or by making the offender stand in Sunday school. If there has been a gross infringement, the offender may have to stand in the back during the evening service or, for a worse offense, he may have to stand in the back during the Sunday service. For a still worse offense, he may have to stand at the front of the church. This is considered too harsh a punishment for a girl to endure. The attitude toward the punishment is revealed by an older Hutterite who said, with intense emotion, "He needs to stand there like a 'dumb ox,' in front of everybody." The implication was that the erring one would really feel how stupid he had been. An offense for which a boy might have to stand in Sunday school would be having his hair cut too short or attending a movie. A boy might stand in church for sneaking off the colony and being arrested by the local police for drinking with minors. The Sunday-school program does not let the young people forget their spiritual status as children who must memorize and recite their lessons correctly at the bidding of their teacher. By physically grouping them with the young children and continuing their ritualistic participation with the school children for religious training, those who are not yet baptized are constantly reminded that they are not yet fully adult. Each week, in effect, they are told, "Although you may look like an adult and work like an adult, really, you are still a child." Sunday school is a place where behavior problems that may arise among the young people can be discussed. The push of society is toward adulthood, even while recognizing and tolerating the childish elements that still remain.

During the work day, both the boys and the girls work with and under the direction of older colony members, but they also work in peer groups: all the young men hay together, all the young girls paint together. Work and social life intermingle, for as they work together they talk together, especially when beyond the hearing of adults. It is considered a privilege to go to other colonies to work, because here too the young people work with their peers and establish new friendships. After supper the peer group expands to include both sexes and visitors of the same age group from other colonies. Frequently there is no one in his home colony whom a young person can date, because all those his own age are his first cousins; or if there are some to whom he is more distantly related, they may be "going with" someone else.

The intermingling of work, visiting, and dating, is illustrated in the following account. On Friday afternoon a truck arrived from a colony in a neighboring province. The driver was a young unmarried man who was related to this colony by marriage and was known to be dating one of the girls in the host colony. She did not know in advance that he was coming nor did he know he would be chosen to drive the truck until he was told to get ready. With him, riding inside was an older woman who was the sister of one of the older men in the host colony. The woman had her youngest child with her, a boy of about nine. She had come to visit her brother. Eight girls came along in the back of the truck, because the host colony had a bumper crop of cucumbers.

The truck arrived too late that Friday to do any picking. Saturday morning the girls picked, but Saturday afternoon everyone needed to prepare for Sunday, so no more cucumbers were gathered. Sunday the young people were neither allowed to pick vegetables nor to travel; they stayed over Sunday. Early Monday morning they picked more cucumbers before the truck left. The visit enabled the girls to work and also to spend the entire week-end visiting. The girls of the host colony were with the visitors constantly, helping them pick, pack, and showing them around. Who was to stay with whom was easily decided, for each girl stayed with her closest relative. Early in the evening all the young people gathered together and sang western songs, usually ones with a plaintive tune and many verses telling some sad tale of worldly life; an orphan with no one to care for her, a mother whose only child is killed after her husband has deserted her, a dutiful son killed in a war. Later in the evening the young people separated into couples. The boys in the host colony decided who would date whom. The oldest boy has first choice and the others decide in order of age. If a boy has a girl friend at another colony, he will indicate that he does not want to date. If it is known that one couple is going together, this is, of course, accepted. If a visiting girl does not want to date, or does not want to date a specific person, she says "No." If boys and girls do not know one another, they tend to pair up by age. The younger boys at the host colony will be included in the group of young people but probably will not have a date, for the older boys will have claimed all the visiting girls.

Dating generally begins when the child becomes a young person. Occasionally and surreptitiously it may begin earlier in colonies where there are children more distantly related than first cousins. Some parents do not allow their daughters to date until they are seventeen or eighteen or even older. There is a considerable individual variation among the young people; some date whenever and whomever they may, others go only with one person for a long period of time. It is not unusual for courtship to last five or six years. Whether or not a young person is dating, he is included in the mixed peer group activity and remains absolutely loyal to the group.

The peer group is of supreme importance, for within it virtually all the young person's social life takes place and many of his working hours are spent with his same-sex peer group. The group demands absolute loyalty. Anyone who transgresses by not supporting the group or by talking about their activities or plans is completely ostracized. Good times are planned, such as picking and boiling corn in the kitchen to eat after dark, and the ostracized one does not hear about it until afterwards. In large colonies or among the young people of neighboring colonies, exclusive cliques may develop, but they present a united front to all those who are not in their age set. The adults conveniently "do not see" the mild transgressions, and the young people enjoy the thrill of semiforbidden behavior and of escape from adult surveillance.

The "foolish years" are a time for trying the boundaries. The young person will eventually grow to the point where he will reject the world and choose the colony way of life, but during this time there is some flirtation with the world, some learning about that which will be rejected. Most young people have photographs taken (no one should make a graven image), many have their own cameras. Some boys own small transistor radios on which they listen to western songs and from which they memorize the songs, some trap during the winter and sell the furs or moonlight to earn extra money, quite a few of the boys own wristwatches and occasionally boys smoke secretly. The girls have colored nail polish, and may use it to paint their toenails which are hidden under heavy black laced shoes. They have perfume, dime-store jewelry, and sometimes fancy underwear. The in-between years are a period of limited self-realization. In extreme cases a young man may leave the colony for a few weeks, several months, or even for a couple of years. He is a "tourist" in the outside world, he learns about it, earns some money, but always plans to return to the colony to marry and raise his family.

There is a tendency, more pronounced among the girls, to create a secret world. As long as the make-believe does not interfere with the young peoples' work and is not flaunted, adults tacitly accept it and, remembering their own youth, are tolerant. Sometimes the secret world is confined to a locked wooden chest; sometimes a corner of the attic is made into a personal microcosm. Here are stored bits of the temporal world, photographs, sheet music, suntan lotion, and souvenirs. These artifacts represent, however meagerly, what the individual has the freedom to pursue or the freedom to renounce. They represent the world outside the colony; they represent the self in its indulgent, vanity-pleasing aspects. As the individual matures and measures these trinkets and indulgences against the full life around him, and as he participates more completely in this very real and very busy life, he generally finds the satisfactions received from active participation in the colony far outweigh those of self-development. His self-image requires colony identification.

During much of this transitional period, the young person is measuring himself first as a member of his peer group and then as a colony member. He generally has a considerable interest in the outside world that eventually helps him to understand better what it means to be a Hutterite. His status within his nuclear family is quite high during this period, before he will begin a family of his own. During the last year or so of his status as a young person, he is expected to show by his works, in other words by his daily behavior, that he can adhere to the rules of the colony. When he has displayed by his actions, and knows with his heart and mind that he cannot continue as an irresponsible child, he willingly and humbly requests baptism that he may become a true member of the colony. The goal of the Hutterite system of child rearing has then been achieved.

Baptism

Hutterite ritual prepares the individual for two important rites of passage. The first is baptism, the second death. Baptism is essential for adult participation in the ritual of daily life; death leads the true Hutterite into life everlasting. There are close parallels between baptism and death. In order to be baptized, the "old man," the natural man, must die so the "spiritual man" may be born; the human body must die, in order that the spiritual man may be released into eternal life. Both rites of passage stress death as an essential step to life.

Every Hutterite has a birthright place in his colony that insures his constant care. Should he die during childhood before he reaches the age of discretion, he is assured a place in heaven. When the individual is able to differentiate between good and evil he becomes responsible for making the correct choice. He requests baptism and, after a period of instruction, is initiated into church membership. He retains this membership until death, unless some major transgression of the rules should cause the church to exclude him either temporarily or permanently.

Baptism is equated with submission to the church. The applicant must desire "to yield himself to God with all his heart and all his soul and all his members . . . to live no more to himself" (Rideman 1950:79). A present-day Hutterite bishop writes: "He who will not be steadfast with the godly, to suffer the evil as well as the good, and accept all as good, however the Lord may direct, let him remain away. . . . We desire to persuade no man with smooth words. It is not a matter of human compulsion or necessity, for God wants voluntary service" (Hofer 1955: 24–25). A seventeen-year-old Hutterite girl said, "Before you ask to be baptized you know here [pointing to her heart] and here [pointing to her head] that you can't live any longer without it."

Since babyhood, the Hutterite has been socialized to believe that the collective unit is more important than the separate individuals who make up the group. From the time he was seven weeks old he has learned to fit into a group pattern, and he has been treated as a member of a group rather than as an individual. He first identified with his nuclear family, then with his peer group in school and, finally, with his postschool work group. With each successive stage of development the number of people above him has decreased, and the number below him has increased. He has been taught to serve and obey those above him, and to care for and direct those below him. Within his peer group there is some competition (primarily in work performance), but there is strong support, especially in the face of threat from the outside. Now, when he requests baptism, the whole colony becomes, in effect, his peer group. From the time he has known what a comb is used for, the Hutterite child has been taught to obey. "When children from little up are used to obeying their God-fearing parents," says a Hutterite sermon, "then they will have formed a habit of obeying, and it will be a lot easier for them to obey Jesus Christ."

Girls are about nineteen or twenty and boys between twenty and twenty-six when baptized. One preacher explained that even the state does not allow people under twenty-one to vote, and baptism, being a much more important decision than voting should not be undertaken much before twenty-one and perhaps a little later. Another preacher explained, "Christ was baptized at thirty years of age and so if someone does not desire this until twenty-five or thirty years of age we must be patient with such a person. Some people don't seem to understand until they are quite old—but it's not right to make God wait so long." The goal of child rearing among the Hutterites is the individual's voluntary decision to submit himself to the *Gemein*. All the child's life has been, in effect, a preparation for this major rite of passage.

"He who is to be baptized must first request, ask for, and desire it," wrote Rideman in the sixteenth century. In all colonies the individual makes the request to be baptized in a highly stylized form and generally with the support of a peer group. The colony members decide months in advance, often a year or so in advance, that they will have baptism that particular year, and generally it is known by everyone in the colony who is going to request baptism and who will wait. There may be a question about one or two of the candidates, but these too are discussed so that by the time the young people are ready to submit formal application the colony is in agreement; quite often a young person is advised not to request baptism yet. The instruction period lasts six to eight weeks during which time and candidates are admonished as a group on Sunday afternoon for about two and a half to three hours. Baptism is held every year in *Schmiedehof,* but many colonies baptize only once every two to five years. If there is only one person who needs baptism he may be taken to a neighboring colony to join their group of applicants for instruction. Normally the baptismal service takes place on Palm Sunday, but also occasionally at Pentecost.

The Hutterites teach that right belief leads to right behavior. Thus it is not enough for the young Hutterite to profess his belief verbally; he must also show in his everyday actions the fruit of this belief. His behavior in all areas must be acceptable to the church community. There is a greater emphasis on correct acting than on correct thinking. The catechism asks, "What is the inner shame?" The reply is "When a man has sinful thoughts, which he should dispose of." It asks, "What is sin?" The answer is "The transgression of the law." Wrong thinking is bad, but wrong behavior is sin.

The whole colony cooperates in admonishing, punishing, and forgiving its members. During the period of instruction for baptism, the applicants are carefully watched, for they must demonstrate that they have really humbled themselves, are devoid of self-will, and are completely obedient to the community. In the course of instruction in one colony, the applicants formally request baptism thirty-six times. Each must know that he really desires and is ready to make the greatest commitment of his life.

The baptismal ceremony consists of two parts: Saturday afternoon the candidates are examined about their belief, and are asked as many as twenty-five questions. On Sunday afternoon the candidates are baptized. The preacher places his hands on each applicant, while his assistant pours a small quantity of pure water on the head of each. The preacher offers a prayer that they may be preserved in piety and faith until death.

One adolescent Hutterite girl explained, "When you are baptized then you are really old." The German school mother said of the changes baptism required, "Up until this time, his will was not renounced, now it is; and he's obligated to report anything he sees that is not correct in the colony. He must always speak to such a one before he reports. He has restricted himself." In other words, baptism signified the voluntary acceptance of responsibility for the actions of everyone in the colony. It signifies the internalization of Hutterite values.

After baptism, men are given voting privileges and are eligible for more responsible work assignments. There is no change in the work status of the girls. Both men and women who have been baptized may attend weddings and funerals in other colonies. It is customary for the newly baptized young people to travel to other colonies. Baptism is assumed to be the first step leading to marriage, for baptism must precede marriage as one's commitment to God takes precedence over one's commitment to one's spouse. Ideally the time between baptism and marriage is not long, especially for a young man. It is said that an unmarried man is like a garden that has no fence. "He needs a wife and family to protect him." Defection is less likely to occur after marriage. Often a young man is not baptized until he is contemplating marriage. A young girl who is seriously interested in a young man may be persuaded to wait to be baptized, and this often has the effect of postponing her marriage for several years.

The baptized but yet unmarried person is on the fringe of the young people's social life. He participates in the pattern of dating, but has little to do with the testing of the boundaries or the tasting of the world that is characteristic of this age. He has made his choice and has, to a considerable extent, lost interest in the material trinkets of a rejected way of life. He prefers to use his energies to help his colony succeed.

With baptism, the relationship between the parents and the child becomes closer. The child has become a member of the colony and his parents treat him as a colony member as well as an offspring. They, with the help of God and the colony, have accomplished their task of raising this child, "in the nurture and admonition of the Lord," and now that he has become a part of the *Gemein* they can enjoy him. The children have become the spiritual brothers and sisters of their parents and they can work together almost as peers and can identify more closely with one another. The child remains emotionally dependent on his parents, but there is no longer the sharp division between parent and child that existed earlier. Within the hierarchical power structure of the colony, the sons tend to cooperate closely with their father and with their biological brothers, and in a highly integrated colony, these patterns are extended to include all the baptized men of the colony.

Marriage and Adulthood

Courtship among the Hutterites is usually long, from about two to six years. There is a great deal of variation in the degree of secretiveness maintained and in

the opportunity couples have to visit each other. Some couples are known by everyone to be going together. This is advantageous because they are more likely to be
able to visit one another when the truck goes between colonies on business. Other
couples may tell no one other than their families until about two weeks before they
hope to be married. Very frequently more than one couple is married at the same
time, sometimes as many as five. Marriages may not take place immediately before
Christmas or from the period just before Easter until after Pentecost.

The description of a specific wedding will illustrate the pattern of festivities.
The couple had known each other for at least five years, but had only visited one
another about five or six times. The bride had never been to the groom's colony.
They were informally engaged for six months before the wedding, but the "news
was not out" until about two weeks before the groom came to ask for her hand. The
young man asked his parents if he could be married; then he asked the first preacher
who, in turn, asked the council. He was given the consent of his colony and a letter
to the preacher of the girl's colony, stating that he had colony permission for the
marriage. The groom arrived at the bride's colony with his father and his grandfather on a Wednesday. They spoke to the first preacher of the bride's colony who
asked the members of the council if they would let her go. In the evening the
groom and his father went to the parents of the bride, and the groom's father asked
her parents if they would give their daughter to his son. Her parents gave their consent and then asked their daughter if she accepted the young man. She consented
and changed to a new dark purple dress, and everyone of church age in the colony
proceeded to the church. Here formal engagement vows were taken. Among other
questions, the preacher asked the young man: "Do you desire to go before her in
such a way that she finds in you a mirror and an example of honesty and will be led
to the Lord through you, so that you may live together as Christians, one being of
benefit to the other?" The preacher added, "Marriage has its share of grief and not
every day is filled with happiness, but brings suffering too, as the women are the
weaker ones." To the young woman is said, "Since God has ordained that the husband is and should be the head of the wife, I ask you: Do you wish to obey him in
all right and godly things as it is the duty of the wife, so that you can serve each
other in Godliness?" The service lasted approximately fifteen minutes. After the
service, the couple returned to the girl's apartment. The suitor took a pitcher of
wine and the bride-to-be a tray with eight tiny tumblers, and they served her parents
and his father. Then they proceeded to every house, serving the people and receiving good wishes for their married life. Everyone including the school children went
into the kitchen. The tables had been arranged around the room; the engaged couple sat in the center opposite the door and all the others sat around the edges of the
dining room, singing wedding songs and enjoying a lunch. About 10:00 P.M. the
children and old folk went to bed, and the young people adjourned to one of the
extra rooms in the colony. There the singing went on until about midnight. Thursday night there was another *hulba* (festivity) in the kitchen. Friday the couple left
for the young man's colony. Before they left, all the colony members brought small
presents to the bride-to-be: a bedspread, a table lamp, flatware, an alarm clock, an
electric steam iron, plates, towels, soap, a small table, a half slip, a bread box, a laundry basket, a wastepaper basket, and pillow cases. In addition, she received from the

colony extra cloth for dresses, bedding, and a new electric sewing machine. The colony contributions are not considered to be gifts; they have been earned and are due the engaged girl. Three panel trucks took the girl, her belongings, and her relatives to her fiancé's colony. Even though the father of the groom-to-be had telephoned the householder that a wedding was to be arranged, as soon as the truck arrived, the young people of the colony eagerly opened the doors, inquiring if the young man had been successful in getting his bride. The young girl in question was introduced to the young girls of the colony and everyone went to the kitchen where they were all fed supper and had another evening of singing.

On Sunday morning the wedding took place. The bride wore a dress of blue brocade. Her uncle, who is the first preacher in her colony, had the major part of the service and asked the couple almost exactly the same questions they had been asked at the engagement. After the vows, he blessed them, saying, "We herewith bear witness that you marry each other as God-fearing partners according to the order of God and the example of the forefathers and with the knowledge and counsel of the elders of the whole congregation." After the service there was a wedding meal complete with wedding cake, followed by more singing and another snack. Everyone was given a bag of nuts and sweets to take home. After supper one wedding song was sung. The bridal couple visited with the guests during the evening and finally returned to their own room.

Everyone in the colony looks forward to a wedding. It is the happiest gathering of relatives and friends. Former acquaintances can be renewed, relatives can visit with one another, and young people can court. A wedding is a lot of work for the colony and quite an expense. This is one of the reasons given for encouraging double or triple weddings. But it is also because the brides and grooms enjoy a joint wedding and like to do things in groups. The more couples being married, the more guests will be present and the bigger and gayer the crowd. Weddings function to formalize marriage, to make explicit the relationship of the two colonies involved, to bring relatives together, and to strengthen intercolony ties by pleasant visitation. In the socialization of the individual the wedding is a rite of passage to a new stage of life. In contrast to the long, rigorous preparation for baptism, the preparation for marriage is incidental. The adjustment for the groom is minimized. Often he retains his former room in or adjoining his parents' apartment. His brothers move out and his bride moves in. There is no change in his pattern of work except that he becomes eligible for a more responsible position.

For the bride marriage involves a tremendous adjustment. Generally she moves into a new colony, sometimes a colony she has never visited. She leaves her parents, her siblings, and her peer group. Her work patterns change both because she is now married, but also because she is in a different colony where things are done slightly differently. She is, however, given a niche, for the bedroom is referred to as her room: not their room or his room. The bedroom belongs to the wife and mother, the colony to the husband. She is under the direction of her mother-in-law. The husband is in an emotionally strong position. All his primary ties are maintained and, in addition, he has a wife. Every detail of the colony is familiar to him. He knows a great deal about all aspects of the colony and about every member of the colony. In contrast, his wife is in a vulnerable position. She has only her husband to turn to for support and information. The marriage patterns function to support the husband in his dominant position and to emphasize the dependence of the wife.

Education for pregnancy and childbirth is even more rudimentary than formal preparation for marriage. Pregnancy and childbirth are rarely discussed among Hutterite women and never in the presence of girls. Daughters are generally living in another colony when they become pregnant and their mothers do not come to care for them until after the baby has been born. There is no recognition of pregnancy (other than her husband growing a beard during the latter part of her first pregnancy), no change in work schedule, no special knowledge that is imparted to the gravid woman. Most young women have been told practically nothing about the details of labor. However, the girls have grown up in a society with a high birth rate and a positive attitude toward babies. They have been taught to ignore discomfort, and to accept without complaint whatever is to be their lot. The attitude toward pregnancy and childbirth appears to be one of passive acceptance, or when there are already many children, of mild annoyance. In contrast, the attitude toward the delivered baby is one of positive enjoyment.

Constant reinforcement is necessary for the adult person. The Sunday and daily services are part of the continuous formal socialization of the individual. These group ceremonials have both the functions of teaching the young and of reinforcing the beliefs the adults have internalized. All adults must set a good example at all times. In virtually all his activities, the adult Hutterite functions within the age-set hierarchy and participates primarily in his same sex group. On the job both the adult man and woman are constantly participating in decision making that must follow a pattern, enabling the group to reach consensus. One individual does not persuade the group or put forward his arguments; rather a subject is discussed in such a way that most of the individuals have participated and the final decision is indeed a group decision. The very youngest adult men may still be in hard labor positions, those older are in leadership positions such as departmental heads of various economic enterprises, while the oldest men tend to be in executive positions, that is, on the council. Although there may be muted competition within the group for various positions and fairly obvious competition among the different positions to demonstrate how efficiently an operation can be run, the general pattern is for the individual to progress to higher status positions as he grows older. Within the well-integrated colony there are no contests for place (place is determined by age and sex) or for office (office is determined by vote of the *Gemein*). Hutterites do not feel comfortable with the satisfaction that might come from a position of dominance. They have been well socialized against self-assertion. "One must not take upon himself" but must "wait until God chooseth." (Rideman 1950:80). Power, authority, and influence are diffused among the mature men and expressed as group action. Biological brothers cooperate closely among themselves and with their father in all areas of life, but this is not necessarily disruptive to the colony as a whole. These men have learned to handle their emotions of jealousy and competition and to work smoothly together, counting on one another's strengths and compensating for one another's weaknesses. The early identification and constant interaction with the peer group have taught the men to work well with those who will be their spiritual brothers. One's spiritual brother is in no sense an outsider or an "other." One's emotion towards him are somewhat less intense than towards one's biological brother and therefore easier to control. In making colony decisions, brother groups are functional. News spreads informally among biological brothers, and among themselves they argue noisily and freely the relative merits of various alternatives. Quite often consensus is reached within one family of brothers. When members of one set of brothers interact with members of another, the problem can be further discussed, with each individual knowing that he has some support. Often a decision can be reached fairly quickly. The men have been socialized against stubbornness, and have been taught that when a colony issue reaches the voting stage, the church members must vote as spiritual brothers rather than as biological brothers. Biological brothers function as a closely cooperating subgroup within the larger cooperative.

Social controls are based primarily on the individual's fear of rejection. The adult Hutterite has identified with the group, and for his own self-esteem, he needs full acceptance by the group. Even an indication of possible rejection is threatening to him. The simplest form of social control is admonishing by a brother. If there is a misunderstanding between individuals, it is their responsibility to straighten it out. If one person is troubled about the conduct of another, he should go and speak to that person. If this is of no avail, the preacher will speak privately to the erring member. Should the transgressor still not desist, the erring individual may be spoken to privately by one or more members, or he may be asked to come before the council and the *Gemein*. If it is a problem that involves a large proportion of the baptized men, such as drinking, several outside preachers may be asked to step in and help. The preachers obtain all the pertinent facts, admonish the offenders in the presence of all baptized men, and then call on them to repent and seal the repentance by each man shaking hands with every other man. There are various degrees of formal punishment for different transgressions. "Pardon" is the least stringent form of punishment for the baptized person. The offender must stand in church after the sermon and confess his fault to the church for "whatsoever she (the Church) forgiveth is forgiven here and in eternity." (Rideman 1950:44). "Discord" is an offense demanding more serious punishment for the baptized member. The offender may be guilty of instigating quarrels or fomenting disunity in the group. He is severed (banned) from church membership temporarily, for a few days or a week, during which time he may not shake hands or eat with other members. He eats alone after the adult meal. Because the Hutterites believe that a person who dies outside the church must spend eternity burning in hell and that his life has been wasted, they take great pains to protect the vulnerable person from any physical danger and from accidental death. He is believed to be repentant when he shows it in his attitude and his behavior, when "one senseth that the Lord hath again drawn nigh to him, been gracious to and accepted him." (Rideman 1950:133). Should an offender continue in his waywardness, as in the case of persistent drunkenness, the ban may continue indefinitely. Some of these persons eventually leave the colony. Other than the shedding of blood, which the church cannot forgive, the worst possible offense is desertion of the colony. The members entertain no hope of heaven for anyone who dies outside the church.

A successfully socialized adult Hutterite gets along well with others, is cheerful, and has a kind word for everyone. He is submissive and obedient to the rules and regulations of the colony. He is a hard, responsible worker. A woman may be praised by the statement, "She is always first in the kitchen when the work bell rings." An adult Hutterite must never display anger or hostility or precipitate quarrels. Intensity and imagination are not admired as are a quiet willingness coupled with hard work. The constant pruning that adapts each individual to the group leads to a minimizing of differences and the muting of emotional expression. The subduing of individuality is implied in the Hutterite saying, "No man with rights has a right to all of his rights." The elimination of extremes and the imposition of a strict order enable the members to find satisfaction in the "narrow way" that leads to salvation.

Aging

Hutterites age early. A man of forty said, "We older people let the young do the hard work. A seventy-year-old person is considered to be extremely old. He looks old and acts old. The formal aspect of the culture requires that the aged be respected by all younger persons. There is no inducement to act younger than one's age. The pressure to move upward in the age set is invariably present from kindergarten through the last rite of passage, death. Older people are believed to deserve "rest." In such capacity, they may spend more time traveling, going to town, talking with neighbors and visitors; work is optional.

There is no abrupt age of retirement. Women usually are relieved of their rotating colony job between the ages of forty-five to fifty. Those who hold the positions of head cook or kindergarten mother continue until they are too old to fulfill these responsibilities. Often women are not assigned to the kindergarten until they are in their forties. The first jobs women give up are milking and hoeing. Milking requires early rising and lifting heavy pails. Usually they also give up cooking and baking. If they still enjoy these activities, there is opportunity to help a daughter or a daughter-in-law or to substitute for someone. Most women help with food preparation as long as they are able, for they prefer working to sitting alone in their apartment. When an older person loses a spouse, he often moves in with one of the children or into a room adjacent to their home. A grandchild is assigned to run errands or even to sleep with the older person to keep him company during the night. When they become too old or weak to go to the dining room, meals are carried to apartments of the elderly. Older women who are doing very little colony work will help look after the babies who are brought to their apartment.

Retirement among men is more difficult to manage than among women. The problem is to persuade an older man to give up his position if he wants to keep it. An elderly man is frequently removed from a foreman position and is simultaneously elected to a position on the council if he is not already on the council. In this way his conservative influence is constructive and the economic development that requires constant change is put into the hands of a younger, more adventurous man. Once elected to the colony council, a man virtually always remains on the council. When a preacher begins to hestitate or to forget his long prayer, it is recognized that he must give up some of his responsibility. Often the most difficult person to relieve is the householder. For although conservatism may be an advantage in a preacher, it is a disadvantage in the householder; the colony must be willing to risk capital and remain dynamic in its economic practices. Several years may be required to change householders. Gentle persuasion, rather than coercive measures, is the rule.

Older people who are ill go to physicians in nearby towns. Good health is important to the Hutterites, but medical practices are influenced by the world view that the body is important only insofar as it houses the eternal soul. Hutterites go to accredited physicians near their colonies, and often go to great lengths and expense to obtain special treatment for chronically ill members. The best physicians, usually from large urban centers, are often unavailable to them on account of the great distances involved. Their rather famed sixteenth-century tradition of healing, bathing, and surgery (Friedmann 1961:126) has been greatly altered, but their own contemporary practitioners include midwives, masseurs, and bone-setters. Physicians who see Hutterite patients report that headaches, constipation, and a variety of neurotic difficulties constitute the major complains. In an intensive study of their mental health, Eaton and Weil (1954) concluded that the group was well poised and extraordinarily free from the symptoms of mental illness in spite of disease patterns that were influenced by the culture. The chances of recovery from mental illness, including an affliction known as *Anfechtung* (a chronic state of deep depression and doubt) are excellent (Kaplan and Plaut 1956:67). The disposition to use herbs and teas, as well as the disposition to practice modern preventive medicine, varies greatly along family lines. Poor health is disruptive when it prevents a person from performing his work. Women especially entertain fears of losing their health, for with the loss of ability to work there is loss of status. A sick person, unless he is old, is not given much personal attention and often must take care of himself.

Old people constitute a small minority of the population of any colony because of the large number of children found in all colonies. The aged are not segregated into a single colony but live with their closest kin. Of a total Hutterite population of 8542 in 1950, only twenty-five persons were seventy-five or over (Eaton and Mayer 1954:9). Life expectancy is believed to be not strikingly different from the United States population, except that men tend to live longer than women. The psychology of aging and preparation for death is unique and consistent with the culture. Old people are not forced to exert themselves nor are they denied active participation. They can reduce their work load without fear of losing status or of being excluded from the policy-forming group, the *Gemein*. The contribution of an old man to the colony is limited economically, but his strong identity with the tradition of the colony and the respect he gets from young members is a stabilizing influence. Communication between old and young is maintained, and old persons can keep abreast of the changing times as much as they wish. Old age is not a period of loneliness and isolation or of economic deprivation. Occasionally the oldest active man cuts the fresh loaves of bread and keeps every table in the communal dining hall supplied with bread. He is given the honorable position of *Brotschneider* (bread cutter). He may eat later with the head cook and the householder. In this capacity the father-figure as provider is symbolically activated in the life of the community.

Death

Death to the Hutterites is the termination of the earthly struggle and the beginning of paradise for those who have lived faithfully. In expectation of heaven, all of life has been preparation for death. Life span, from beginning to ending, is determined by God. The time of death is believed to be controlled only by God. After the assassination of President John F. Kennedy, one Hutterite informant said, "His time was up, and no matter where he would have been, up in an airplane or down on the ground, God called his number and he had to go." A basic attitude is expressed in a Hutterite proverb: "When a person comes to his end, great faith is more important than great possessions."

One day the women were discussing a tragedy that occurred to a neighbor's boy. The lad darted in front of the school bus to cross the road and was killed instantly. "Do you think the boy was just trying to get it over with before he grew up?" asked one. "No. I believe his number was called and he had to die," said another. "It's not so bad when a child dies like that, but we don't like to see a grown-up die so suddenly. We prefer slow deaths, not sudden deaths. The people who want to die suddenly are the outsiders," said a preacher. "They want death to be unexpected so that they do not need to think about it, for they don't want to face the future. We prefer a slow death, even if it involves pain, so that we are sure to have plenty of time to consider eternity and to confess and make everything right."

Children who die in their innocence are envied by adults. They are presumed to be innocent, and are thus spared the temptations and the life-long struggles of self denial. One day little Matilda was playing near a canal and fell into the water. Her older brother rescued her and brought her home, a distance of about one-half mile, on the two-wheeled play cart. She shivered with cold, developed a fever the following day, and pneumonia was predicted. Commenting on the mishap, the preacher said, "If Matilda would have drowned, she surely would have been an angel." An older man said, "When these little ones die we know they are in heaven, but we never know what will happen to them if they grow up. I sure wish I would have died when I was a kid."

The deathbed account of a seventy-five-year-old grandmother, written by one of her children says:

> In her last years our mother occupied herself with knitting and sewing. Her songbook was always open and she learned many lovely songs and verses by heart. Her spirit was directed heavenward and her spirit was quickened through these songs. Her final sickness lasted four weeks and during this time she was examined concerning her life and her goals. All this time she had great difficulty breathing and sleeping and was always conscious and, like the apostle Paul, wanted to go home to be with Christ. She did not wish to recover in this world because her inner pilgrimage had already begun. She looked to her heavenly abode, to which she had already sent many prayers and lovely songs. At her last she told us: "Children, I tell you, live godly, work hard, be faithful, and share with the poor." When she had almost passed on, she asked once more to get up. She stood up straight and then sank into her bed. There she lay, weary and tired of the world. She called her sister to her once more and said, "All, all is finished."

In another deathbed account, a father dying of cancer awoke and began to laugh loudly in the presence of many people (Eaton 1964:94). Asked why he was laughing, he explained that he saw "the heavenly pleasure." He said he saw his father, his father-in-law, and other departed relatives. When asked how they were dressed he explained: "All alike, one like the other." He made this comparison: "Take the balls of a ball bearing; they all look the same and, believe me, that is the nicest of it all."

The emotional acceptance of death is supported by many aspects of the culture, specifically by the lack of sentiment attached to property and by adequate provision for all the survivors. Terminal illnesses, rather than sudden or accidental deaths, permit the individual to confess any wrongs, to admonish the children, and to gather the family and close friends, even from distant colonies, about the dying person. The family and the colony is supportive in time of death. Funerals are extremely important occasions, for they permit integration of the society on all levels. Relatives and friends from distant colonies participate in the religious and burial services. Funeral festivities also permit young people to make new friends and look for a prospective mate.

Individuality, which is denied and carefully guarded against throughout life, is completely abolished in the heaven envisioned by the Hutterites. Throughout life there is little speculation about heaven or the kinds of enjoyment or activity anticipated. What matters most is preparation, through submission to the divine and right order. There is to be more perfect communal living in heaven. The communal emphasis is inculcated through vigorous teaching from a young age and is reinforced in the visions of the dying. The beliefs and practices so consistently taught during life provide meaning for the members, so that in the final rite of passage there is fulfillment.

5

Disruption Patterns

INSIGHT INTO A CULTURE is gained not only by observing the society when it is highly integrated, but also by observing the culture when there are various threats and unusual problems such as war and population decline. In this chapter the major disruptive patterns in Hutterite society, both internal and external, will be described. Important external threats in their history have been war, hostile outsiders, land restrictions and, to a lesser degree, the public school.

War

War and its hardships have on several occasions almost extinguished the Hutterites. In their world view Hutterites are not opposed to governments, and their leaders often say, "A poor government is better than none at all." Hutterites respect order and discipline in governments or in the military. But for them, as with the Anabaptists, government is believed to be necessary for the carnal order. Christians, they say, should obey government in all matters that do not violate biblical teaching but should not involve themselves in the administration of government or holding public office. "Christ refused to defend himself and set an example for his followers." The Old Testament law of retribution was repealed, and for the Hutterites warfare must come to an end with the establishment of the true church of Christ. The nature of Hutterite conflict with government has varied with the particular issue in question and the circumstances. Their European experience contrasts sharply with that in North America.

Their early period of prosperity in Moravia ended in 1593 when Turkey and Austria went to war (Friedmann 1961). Soldiers were billeted in the colonies and possessions and livestock were taken. When the Turks invaded Moravia in 1605 they destroyed sixteen colonies and carried away several hundred captives. The Thirty Years' War came in 1618, ending in dire consequences for all of Europe. Moravia joined the Protestant cause, only to be invaded by the Hapsburg army in 1619, which destroyed many colonies. By 1620 the Roman Catholic forces suppressed all other religions in the Hapsburg empire. The soldiers raided the colonies and, after torture and trickery at the hands of Austrian and Polish soldiers, the last twenty-four Moravian Hutterite colonies were abandoned.

The remaining devout Hutterites who did not accept Catholicism escaped to Slovakia and located there in fifteen colonies, which survived until 1760. Life in Slovakia was hard and the Jesuits, with encouragement from Maria Theresa, succeeded in confiscating Hutterite literature and in converting members and baptizing their infants into the Catholic church. Those who turned Catholic for several generations secretly adhered to the communal faith and practice. With forced conversion in Slovakia, a small group of nineteen persons became the thread to survival. A group of fifty-six Lutheran refugees joined the remnant of nineteen remaining Hutterites in 1756 in the small country of Transylvania. Just as the Hutterite children were about to be placed in a Catholic orphanage, the group fled by night to Wallachia. Here they were soon caught up in a war between Russia and Turkey.

An appeal to a Russian general led to an invitation to settle in the Ukraine. They were granted religious freedom and liberal terms of settlement in line with the policies of Catherine the Great toward other German immigrants. A hundred years later, when conscription was enacted, they migrated to the United States. Thus, in Europe, the external pressures of war, plunder, coercion, or forced colonization were direct and disruptive. Only by successive appeals to governments more sympathetic than their own were they able to survive. A remnant remained loyal to the core of their faith. The chronicle of the early period (Wolkan 1923), still well-preserved in a contemporary colony, contains a list of 2175 persons who died a martyr's death during this extended period.

In America, war and the fear of conscription scattered the original settlements in South Dakota to other states and to the provinces of Canada. With the outbreak of the Spanish-American War, the *Dariusleut* sought to move to Canada to escape possible military conscription. The privileges asked of the Canadian government were granted in 1899. One colony moved to Canada but later returned because of poor farming conditions.

The first Hutterite confrontation with conscription in North America occurred during World War I. As American patriotism swept the country, and the midwestern states in particular, the Hutterites suffered harassment from their neighbors. Many Americans thought they were pro-German and therefore enemies. Young Hutterites were drafted and ordered to report for army duty. They refused army service and were segregated in army barracks and guard houses. The law made no provision for sincere conscientious objectors. Hutterite leaders sent a special appeal to President Woodrow Wilson. Three Hutterite representatives went to Washington to seek clarification of a law that provided for noncombatant service for religious objectors under terms to be prescribed by the President, and to obtain permission for their young men to move with the colonies to Canada. They met with little success because Secretary of War, Newton Baker, could not be located. Meanwhile, young men were being court-martialed and placed in barracks and ridiculed at home. The colonies were humiliated and persecuted by unsympathetic neighbors. They would "capture" a young man and forcibly cut his beard into ridiculous shapes. At other times they smuggled glass into the flour mills of Hutterite colonies in order to accuse them of sabotage. The State Council of Defense set as its goal the elimination of all pro-Germanism, prohibiting the use of the German language in public or in churches. The newspapers charged the Hutterites with making wartime profits and for their lack of patriotism. Although Hutterites were willing to give money for relief and offered $10,000 for this purpose, they refused to buy war bonds. A group of patriots visited the Jamesville colony, and without opposition, drove to market a hundred steers and a thousand sheep to force the colony to buy bonds. The packing houses refused to touch the stolen cattle. They were sold for $14,000, or about half their true value. The money was refused by the National War Loan Committee and was then deposited in a local bank in the name of the Hutterites. The Hutterites refused to accept the money. (Later it was accepted by a Canadian land agent as a down payment on land.) The colonies experienced raids from local citizens who confiscated their wine under the Prohibition Act but drank it at the county seat. The ridicule and torture of four drafted men—Jacob Wipf and three Hofer brothers, Joseph, David, and Michael—was one of the most shocking to come out of this country's experience with conscientious objectors (see pages 9–10). The cruel treatment and deaths of Joseph and Michael Hofer seemed to the Hutterites like a return to the sixteenth century. All of these events culminated in a mass migration to Canada beginning in 1918.

The move to Canada did not proceed without difficulties concerning Canadian immigration laws, the sale of land at sacrificial prices, and a final legal suit brought by the state of South Dakota. The State Council of Defense had stipulated that 5 percent of the sale price of Hutterite land be invested in war bonds and that one-half percent be given to the Red Cross. The Hutterites resolved this problem by lowering the selling price by this amount and letting the buyers purchase the bonds. In a final move against the Hutterites, the State Council of Defense persuaded the state attorney general to use legal action to revoke the articles of incorporation of all Hutterite colonies. The contestants in the trial claimed that the Hutterites had amassed a fortune of over a million dollars in assets and "had dedicated none of it to the worship of God according to any religious belief, for they did not even have a church building" (Conkin 1964:62). It was argued that the corporation had abused its articles of incorporation in unauthorized business activities unrelated to religion, and on this basis the state court asked that it dispose of all land and secular assets within ninety days. Noncompliance meant that the corporation would be declared bankrupt and all assets would be seized. Those who advanced the suit described the Hutterites as a menace to society. They resented their use of the German language, the control of the colony over the children, and their restrictions against mingling with outsiders. On the basis of the suit, the State Council of Defense invited individual Hutterites to leave the colonies and to claim their share of colony property. Not a single individual took advantage of the invitation.

With the outbreak of World War II, there were fifty-two colonies in all of North America. Only six colonies were left in the United States, in spite of the efforts of local governments to attract the Hutterites back during the financial depression of the thirties. One colony was in Montana and five were in South Dakota. In both the United States and Canada there was a more tolerant attitude and greater facility for dealing with conscientious objectors than during World War I. The Selective Service Act in the United States provided for the assignment of conscientious objectors to civilian public service workcamps. The cost of the camps was underwritten by various religious groups. Hutterite draftees were assigned to these camps, and the colonies contributed toward the cost of maintenance. Conscientious objectors in Canada were assigned to forestry service. Colonies were required to contribute monthly assessments to the Red Cross for each youth drafted. Many were deferred for farm work. A few Hutterites chose jail. There was a total of 276 Hutterites in both countries who served in workcamps for objectors, and 26 who joined the armed forces. Since World War II, young Hutterite men in the United States are drafted as conscientious objectors and assigned to work in national parks. They receive a maintenance allowance from the government for their living expenses and continue to live communally in a separate house during the two years of government service.

Throughout their history, war has been a significant external force on Hutterite society. In earlier times the society was in danger of complete extinction by genocide and forced assimilation. In more modern days wars have produced migration and have intensified the cleavages between the colonies and the "world." The twentieth-century treatment of the Hutterites is but a subtle continuation of the persecutions that were so widespread in the sixteenth century. The direct hostility of the sixteenth century has been displaced by more humane considerations and a tendency for the dominant society to use more sophisticated (legal) means of expressing intolerance. But to the Hutterite, the modern forms of persecution have a more sinister connotation than the cruel persecution faced by the founding fathers. Renewed intolerances are a constant reminder to the Hutterites that they are completely dependent upon God for security.

Neighboring Practices

The colonies vary in their neighboring practices. The formal pattern forbids certain kinds of associations with non-Hutterites, which, as stated in the chapter on world view, are based upon the doctrine of separation from the world. Of the several colonies studied intensively and of a great many observed, it was found that the most self-sufficient colony tended to display the most conspicuous ethnocentric patterns and had the poorest public relations. The colony with the smallest number of people and most geographically isolated was deliberately cultivating favorable neighborly relations. One colony that is widely reputed for its progressive agricultural practices permits visitors regularly by appointment. After a tour of the grounds, lunch is served, and the visitors are seated in the church for a "lecture" by the preacher, with a discussion period following.

Attitudes toward the Hutterites prior to their moving into a new area are often very different from those that prevail once a colony is established (Serl 1964:111–129). In addition to the fears expressed that a colony would not contribute to the local economy and that land prices would soar upward, there is an element of vexation caused by the Hutterite refusal to mix socially in the local community. Contrary to the fear of local representatives concerned with the economy, Hutterite colonies do make substantial economic contributions to the community as has been observed in Saskatchewan by Bennett (1967). In one small, new colony, in an area where there are no other colonies nearby, formal neighboring relations are excellent. Neighboring farmers come to the colony for specialized tractor-repair services, to get their seeds cleaned in colony plants, and to get machines for spraying their crops. These services, often with additional assistance in planting and haying for neighbors, are more a gesture toward cultivating good neighborly relations than a desired source of colony income. "We have to be accommodating, and we will not turn anybody away," said the colony preacher. The same colony also loaned a young man as a tractor operator to the municipal road building crew for a few weeks.

The large colony, which is often self-sufficient and less sensitive to maintaining good public relations, views its informal relations with neighbors differently. The predicament was well stated by one colony spokesman:

> A good neighbor is one we never see, talk with, or help back and forth, or that never comes on the place. We have a number of neighbors with whom we exchange work and machinery. In winter when the roads are snowed shut we loan our teams to them. The more contact we have with neighbors, the more chance our boys have to buddy up with them. We don't mind favoring them with help, but then they want to favor our boys by taking them into their homes, letting them listen to the radio, television, taking them to shows, and then our colony rules are broken. When we tell our neighbors not to do this, they just get mad, and then there is friction.

Hostile acts from outsiders are taken for granted and are built into the world view. When acts of hostility do occur, they tend to function as a cohesive force, thereby integrating the structural relations of the society. When there is no "persecution" or anti-Hutterite sentiment, Hutterite leaders acknowledge a tendency toward internal disruptive patterns. While publicity is not shunned, favorable articles about them are disturbing to some of the more sensitive leaders who say: "We don't like horn tooting. It makes people jealous, and creates hard feelings." Writers and reporters who visit colonies and hear individuals unburden their grievances often assume naively that relations can be helped by the right kind of publicity. Additional knowledge about Hutterites does not necessarily assure a more tolerant attitude on the part of neighbors (Priestly 1959). Vandalism from unsympathetic gangs is unpredictable and harassing to colonies. "The town people often call to tell us to come and get our geese which are roaming the streets," said a Hutterite woman. "We know who the teen-age boys are that let them loose but their parents don't care and the police do nothing about it." Other intrusions range from broken windows, pilfering tools and supplies, to armed robbery. Cattle thievery occurs in some of the sparsely populated areas of the prairie country. One colony suffered the loss of thirty head of fattened hogs. Pranks such as putting harmful substances into gasoline tanks are disruptive and costly. Some colonies are summoned by telephone in case of fire in nearby small towns; they help willingly. Once when a false alarm was sprung on the colony, a Hutterite told a townsman:

> Don't depend on us anymore in case of fire. The last time a call came at midnight, and the snow and mud was as deep as our butts. We got our trucks and tractors out, spent the whole night until 8:00 in the morning trying to get our vehicles home. Some fool pulled a false alarm.

Hutterite segregation is perhaps one of the most intolerable vices as viewed by the larger society. The many symbolic ways in which Hutterites isolate themselves gives rise to a variety of myths and falsehoods about their communal life. These falsifications eventually are harmful because of the misinformation provided to the public via mass media. In spite of these distortions, the persecution and dislikes expressed by outsiders tend to strengthen the internal cohesion of the colony.

Individual acts and habits, like not leaving a tip in the restaurant or refusing to lead a prayer at a village gathering, do not trouble the conscience of the individual Hutterite. The individual Hutterite involved may not realize that although his own reputation is not at stake, there is frequently a negative reflection on the colony as a whole.

Land Restrictions

Most of the adverse sentiments expressed against Hutterites can be grouped into three categories: dislike of their separatism, charges of unfair economic competition, and fear of unprecedented growth. As long as these negative sentiments operate informally, they present no major threat to the colonies. When they are formalized into laws, they disrupt the expansion patterns of the colonies. Alberta passed a law in 1942 that prevented the sale of land to enemy aliens and Hutterites. In 1947, the colonies were permitted to buy up to 6400 acres of land for each new colony if the site was forty miles from an existing colony. The law was changed in 1960 by eliminating the forty-mile clause and allowing no land sales to Hutterite colonies without hearings before a community property control board and approval by the legislative cabinet. The board must recommend in each instance "whether or not it is in the public interest to permit the sale of land." Since these laws have been in effect, colonies have tended to form new colonies northward in Alberta and to locate in Montana (since 1948) and in Saskatchewan (since 1952).

When the Alberta colonies branched into Saskatchewan, there was immediate local opposition. The provincial government conducted a survey (Canadian Mental Health Association, 1953) to explore the nature and extent of the problem and concluded that the colonies presented no threat and that future clashes could be alleviated by disseminating proper information. A Provincial Committee on Minority Groups established a liaison officer to anticipate and assist in a settlement program which would free communities from the fear of being overrun by the colonies.

There was agitation in Manitoba to introduce legislation to restrict the colonies, but a legislative committee urged that no restrictions be placed on "the fundamental right" to purchase land. All attempts at restrictive laws in Manitoba have failed, but in 1954 the Union of Manitoba Municipalities demanded restrictions on the acreage and location of colonies. This culminated in an agreement in which the Hutterites consented to locate no more than two colonies in large municipalities and one colony in small ones, to limit new colony acreage to 5120, and to keep colonies at least ten miles apart. The threat of legislation affected the land-buying pattern of the colonies, in some instances long before any formal agreements were reached. Some colonies in Manitoba responded to the threat of legislation by establishing daughter colonies in South Dakota.

The effect of Canadian land restrictions on the colonies has been a dispersion of their settlement pattern into wider geographic areas. The older colonies are unaffected except for limitations on additional land purchases. Another effect has been the formalization of the three *Leut* into a single legal entity. With the suggestion and assistance of legal advisors, a charter was granted to the Hutterite colonies of Canada by the Canadian Parliament in 1951. Each *Leut* is defined as a church conference and elects three members to a nine-member board of directors of the Hutterian Brethren Church. The board elects a chairman or senior elder and meets annually, usually with a legal representative. The formal organization allows more unified representation to government, encourages discussion and prediction of problems among the colonies, and encourages greater consensus among the *Leut*. This Canadian charter specifies that property rights belong to the individual colonies, and that disciplinary functions are the prerogative of each *Leut*. The charter protects the colonies against individual members who might desert the colony and make a claim to the corporate property.

After very severe measures to dissolve the colonies during World War I, South Dakota encouraged the colonies to return during the years of the depression. Pacifism was a remote issue, the state needed tax revenues, and local governments needed occupants for deserted farm lands. A law was passed in 1935 enabling the colonies to incorporate with the same tax privileges granted to cooperatives. While they would have to pay local taxes, there would be no state and federal corporate taxes. Many of the colonies returned. But faced with bitter opposition to the colonies, the South Dakota legislature passed a bill in 1955 intended to curb Hutterite expansion. It prohibited expansion by any of the incorporated colonies (fifteen in all) from the purchase or lease of land. The bill did not prevent new colonies from forming. Efforts to introduce restrictive laws in Montana and Minnesota have to date not succeeded.

Individual colonies each face their own local and often very peculiar legal tangles. Although the leaders are adept in cultivating friendships with influential outsiders and of arguing on a practical level, they are also trapped into exorbitant prices and leases that require legal counsel. Virtually all colonies have a legal advisor. The several suits brought by defecting members against colony corporations to obtain a share of the communal assets have all failed. The promise not to lay claim to the colony property is written into the baptismal vow as well as in the articles of incorporation.

The popular impressions that Hutterites would purchase unlimited acres of land to add to their present holdings if they could and that they would exercise no voluntary restraints on either the location or pattern of settlement is not true. That they will eventually outnumber the rural farm population, a fear expressed by some farm organizations, has no empirical basis. The sixty or so colonies in Alberta own or lease on the average about 7200 acres per colony, less than 1 percent of the agricultural lands in the province. The average number of acres owned per person in colonies is 58, while land acreage per person among farm families in Alberta counties where Hutterites live is about 122. Lands owned and operated by South Dakota colonies in 1957 averaged 4640 acres per colony (Riley and Priestly 1959). The farm population displaced by new colonies varies with the area, but is usually minimal. One colony of sixty-five persons acquired 6000 acres of land, and the total population displaced was only six persons. The farms acquired are frequently operated by absentee owners or by aged farmers who wish to retire. Studies of the spending habits of the colonies in comparison with farm families show distinct differences (Bennett 1967). Hutterites spend less per capita for consumer goods than do individual farmers, but the influx of colony population into the area requires expenditures of an amount no less than that spent by the displaced population. The colony consumes large amounts of building supplies and farm equipment.

English School

The English school becomes a disruptive force when social distance is not properly maintained. In its place, the school contributes to colony cohesion. Keeping the English school "in its place" is crucial, for here is where the ideology of the world and of the colony vie for the loyalty of young minds. For the child who has not responded properly to colony indoctrination, the English school can become an important influence leading to possible desertion. In school he can function as an individual and learn about the world outside his colony from his books and from his teacher. Intimacy between teacher and pupil can lead to defection. Friendships can lead to marriages with outsiders, to changes of denominational loyalties, and in some instances, teachers have helped young Hutterites find jobs and leave the colony. Young, single teachers, male or female, are greater risks than older teachers who are married and have children of their own.

All children of school age attend the English school of the colony. The usual pattern is for the colony to supply the building, the heating and the maintenance costs; the school boards selects and pays the salary of the teacher. Many of the more isolated colonies provide a small house, a teacherage, for the teacher and his family. Whether the school is private or public, the curriculum is similar to that which is taught in rural schools. The colony makes a point of not interfering with the living pattern of the teacher who may have a radio, a television, and even a separate mailbox. The home of the teacher becomes an informal source of worldly knowledge to colony people and a source of intrusion if not properly controlled. Teachers who have never taught in a colony are typically given a "lecture" by the preacher at the start or with their first offense. The preacher outlines the colony's expectation of the teacher and sets the limits for practices which are "against our religion." Projected audio-visual material, radios, phonographs, and recorders are forbidden. Since the school building in most colonies becomes a church every evening, the walls must be bare. The blackboard must be erased and pictures must be removed or turned to face the wall at the close of each day of school.

The attitude toward English school is that it is important, especially in the early grades, for all children must know arithmetic and to be able to use the language of the country. One leader said: "We expect our children to learn math, reading, and science as required by the Department of Education. We must learn English to understand the people around us." When asked what is most undesirable about the English school, these answers were given:

When the teacher does not cooperate with the German teacher or the preacher; taking pictures and then distributing them to the children or when dancing is held in the school, as it happened once. Learning the worldly ways would lead to their damnation. The old physiology books were all right, but the modern health books contain too much about dating, sex education, and anatomy.

Aside from acquiring a good knowledge of arithmetic and reading, the only additional goal for Hutterites is that discipline be maintained in the English school. Teachers are expected and often encouraged to "lay down the law," and if they cannot maintain order, they are considered failures. A German teacher advised a new English teacher to "use the willow, for it's the only language they understand." A major complaint of teachers is that the children lack self-discipline and have less respect for an outside teacher than for the colony's German teacher. Greater respect for colony authority can be maintained if the English school also supports the prevailing authoritarian pattern.

The colony will accept the English school complex but restrain and bridle its influence so that it will serve colony ends. The school and teacherage are on the grounds but they are oriented to one side of the colony and can function without major interference. The English school remains emotionally outside the colony. The time patterns, schedule, and colony holidays suggest superior loyalties. The first language learned and the first writing skills acquired are in German. German school is more important in a child's day than English school, as is evident from the time of day in which it meets. German school is held at the start of the day, followed by English school. In effect, the English school is held in the visible presence of the elders, suggested by the council bench in front and the pews in the rear of the school. Thus the English school is encapsulated by the colony pattern, and ideally its influence cannot go beyond the bounds set by the culture.

The influence of the teacher is carefully delineated and maintained on an informal level. While the school board selects, assigns, and pays for the salary of the teacher in most colony schools, the colony has little veto power if the teacher proves unsatisfactory. The teacher is given moral encouragement in ways that aid the colony pattern, mainly as a strong supporter of discipline. The teacher cannot encroach on the child's colony time pattern by asking a child to stay after school, by staying during lunch hour, or by homework assignments. A child may not be punished by depriving him of food. Discussion of a teacher's shortcomings in the presence of children limits her influence. Many of the colonies complain about receiving poor teachers. Indeed some are inferior as teachers, but there is evidence to prove that many of them like to teach in the colonies. Relatively inferior academic teaching is tolerated, and teachers are virtually free from the informal supervision of superiors. There is little pressure from parents for excellence and no parental interference in teaching. Unlike most school teachers who must participate and relate to the wider activities of the community, the teacher in a colony enjoys a type of privacy and freedom from such demands. The formal relations between colony adults and the teacher are cordial and are typically enhanced by gifts of food and help. Low-cost housing and eating privileges are enjoyed by many of these teachers. Those with cooperative attitudes, including the poorer teachers, are more readily absorbed into the environment of the colony than teachers who are truly competent by outside

standards and demand independent thinking of their pupils. Even then, there is little danger that the teacher will become a model for the children. When the teachers
will emulate the colony pattern, in dress or by wearing a beard, disruption patterns
are minimized and the children tend to show greater respect for the teacher.

Attempts by Hutterites in the past to permit their own young men to enter
college and acquire teacher training have largely failed. Through these attempts
some of their young men in both the Ukraine and in North America deserted the
colonies. Although there are presently three college-educated Hutterite men teaching colony English schools, the practice is not favored with unanimity by the leaders. The reason as given by one spokesman is: "It is better to have the worldly
school taught by a worldly person so that we can keep the lines straight."

Laws intended to raise the minimum attendance age, requiring children to
take formal schooling through grades 9 or 10, are now adversely affecting colonies
in some states and provinces. A few Hutterite young people take correspondence
courses from state colleges. Some of the teachers who are assigned to teach in the
colony are willing to tutor or to give instruction beyond the elementary grades.
Exploratory efforts have been made to establish high schools for Hutterite young
people in regions where there are many colonies. All such efforts to take the pupils
from the colony grounds have run counter to Hutterite religion and to their conceptions of child training. The Constitution of the Hutterian Brethren Church (1950)
itself assures every child of an education in skills needed to participate in the adult
community. The consequences of having to attend school beyond the time when he
is accepted as an adult in his society (age fifteen), adversely affects the pupil himself.
At this age young people are given adult work privileges and serve as apprentices
to skilled adults under closely supervised conditions of learning. There is a tendency
for the young Hutterite to feel deprived of his status as a growing person when forced
to attend formal schooling beyond the age required by his culture.

The farther the child goes in school the less he is said to learn. From the
colony's point of view this is correct, for once a child has mastered the basic skills,
much of the rest of the subject matter learned has little relevance to their way of
life. The colony German school teaches the children how to live, and the English
school teaches facts, many of which are of little use to them. German school teaches
proper ritual, the English school teaches worldly knowledge. The schools are clearly
different, and both are regarded as necessary. In the ideal colony there is little
conflict between the two schools, and the normal child receives an integrated learning experience from the viewpoint of his culture.

The disruptive influences described thus far are the major external forces
affecting the colonies. These forces usually have affected the Hutterite culture in
such a way as to strengthen the dedication of its members. The internal patterns of
disruption, affluence and poverty, leadership failure, and defection will now be described.

Affluence and Poverty

The restraints on spending money are clearly delineated in the Hutterite
world view and in their practices. The purpose of all wealth getting, as Rideman

(1950:126) emphasized during the precapitalistic age, is that "Men should labor, working with their hands what is honest that they may have to give to him that needeth." Industriousness and thrift in Hutterite society are means to a style of life pictured by Max Weber (1905) as "ascetic Protestantism" and reformulated by Karl Peter (1965:146) as "a creative and original solution by a *Gemeinschaft*-type of society." The principle of austere consumption is consistent with the "asceticism of the sects" expressed in the refusal to use the law, to swear in courts of justice, to exercise dominion over others, or to take part in war (Troeltsch 1931:331–343). The "narrow path" allows no pursuit of pleasure of its own sake. The modern concept of consumer spending, in contrast with the Hutterite concept of austerity, is based on a different conception of human nature.

The dangers of wealth to Hutterites are primarily two: too much wealth and its adverse influence on a colony, and the temptations of individuals to attain either forbidden commodities or to obtain money by selling colony property. In Europe, money was accumulated and secretly concealed to carry the colonies through critical periods of war, famine, and migration. In North America, the accumulation of money outside of banks is no longer necessary, for banks have proved to be trustworthy. The major function of money is to provide capital for the expansion needed to keep pace with the population growth. Today saving is an institutionalized pattern of all colonies, and the major incentive is to save for the time of branching. A debt-free colony with additional accumulations of money in the bank may spend it for certain extras on farm machinery, tractors, or combines. Added machines tend to make available more positions for younger men who are eligible for the important jobs. But too much money has certain dangers, even for the welfare of the colony. Wealthy colonies tend to obtain more labor-saving commodities, and become the envy of others. This wealth is evidenced by a high degree of mechanization, especially for the work of women, which is usually the last phase of mechanization. Automatic coffee makers and electric floor polishers are presently the exception rather than the rule. A certain amount of consumer goods can be permitted if all share equally, but the danger is in its effect on other colonies and on the individuals themselves whose wants are increased.

A wealthy colony that fails to branch can prove disruptive to the communal way of life. Excessive wealth with unresolved conflicts between informal factions have led to disruption. The reason for not branching is that "the colony becomes too content" with its financially prosperous reputation and with its "luxuries," and there is the added complication of not being able to agree to branch or to resolve the problem of polarization that tends to exist between brother sibships.

Colony wealth tends to have a strange influence on the marriage patterns. With colony affluence, there emerges what Hutterites call "girl power"—the tendency for girls to put off marriage for a number of years to enjoy the conveniences of their own colony. Few girls are fortunate enough to obtain a husband in their own colony since patrilocal rule requires that girls join the colony of their husbands. In a prosperous colony, prestige reaches a point that is rather atypical for the colonies as a whole. The ascetic restraints on consumption are slightly modified to allow for the enjoyment of small luxuries. The girls say, "Why should we get married, we have it so good here." Thus marriage is put off until the late twenties, just short of the "old maid" stage. Were this practice to become widespread, there would likely be a marked decline in the number of children born to a family. A colony of poor reputation, whether through chronic drought or poverty or mismanagement, also has marriage problems. A boy receives no encouragement from his parents to marry a girl who belongs to a colony with a poor standing, and no girl wants to marry into a poor colony.

Individuals who receive private income beyond the small allowances given by the colony are a potential threat to the ideals and practices of the colony. Unauthorized money can be gained privately in several ways, and it is a perpetual problem for the colony leadership. Laboring for a nearby farmer during evening hours is one source of income. Money can be obtained by selling or trading colony property: a duck, turkey, chicken, or a bottle of colony wine, with outsiders. If caught the offender will be punished.

Trapping is a tempting and occasionally a profitable private enterprise. Young boys of school age will trap animals like rabbits that can be caught with little effort, while older boys catch cunning animals and those with the more valuable furs. Groups of school-age boys will assist each other in the activity and will exclude those not of their age or clique. Frozen rabbits can be accumulated in the very cold winter until an older, unbaptized person will take them to town for a cash price. A boy has been known to catch twenty-five rabbits in a single winter. One boy, in one season, trapped 150 muskrats worth $1 each, and he also caught several beavers and mink. Suspicious mail that comes to the colony is opened by the preacher or the parents. To overcome this barrier, boys who are assigned to prolonged jobs away from the colony's buildings, like herding sheep or bulldozing, will rent a personal post-office box. For a nominal fee, a box can be secured in a town not usually patronized by the colony. Furs may be shipped to far away places and through mail order, the boys can secure transistor radios and other private, but forbidden property. The small, secretive adolescent clique can often do an effective business until the subversion is discovered. All of this in the Hutterite view attests to man's natural, selfish desire, and with baptism all such subversiveness must be confessed. The consequences of not confessing all previous sins can be disastrous. One young man who had left the colony and returned in his youth was ordained as preacher. He was liked by most colonists and achieved a very favorable status. After several years, his secret sins began to bother him. He confessed that he had earned money while outside, invested it, and was still collecting interest. He was excommunicated for one week and permanently relieved of his position.

In some of the fringe colonies there is a tendency for young men to accumulate savings with which to buy furnishings for the home after marriage. In still other colonies, savings secured by working for neighbors are accumulated by family heads. This practice leads to serious consequences for the colony as a whole since it introduces differentiation between families. The family-allowance benefits available to family heads in Canada have been disruptive. A monthly allowance of $6 per child under age eighteen can provide a substantial income for a Hutterite family. When some families receive the government benefits and others do not, there is ill feeling between families. A few colonies attempted to control the situation by permitting all families to receive the benefits that go into the colony account. Preacher conferences, however, have ruled that the practice is too disruptive for the welfare of the colonies. At the present time receiving the allowance is outlawed, but offenders are not uniformly dealt with.

To the Hutterites, private wealth is a means by which the individual can be drawn away from his proper spiritual relationship. Luxury must be avoided. The temporal life of the individual must be controlled by the spiritual order. Thus, while material goods are necessary for the body, they must not become the object of pleasure. The human body itself is important only because it houses the soul that is capable of establishing an eternal relationship with God.

Leadership Failure

The prime leadership role in a colony is that of the first preacher. He heads the council, which is responsible for the implementation of policy made by his own colony and by the preacher assembly which ordained him. In decision making his vote counts no more than others, but his recommendations are typically supported. In its routine affairs a colony is autonomous; decisions, once brought to vote, are decided by majority. If there is mismanagement within a colony, any member may "complain to the elders," the preacher assembly. The preachers will conduct an investigation and intervene where a leader is believed to be guilty of misconduct. A preacher can be deposed from office. A preacher who neglects to elect a householder in a colony can be ordered to do so by the preacher assembly. Ideally, every member is led by the Holy Spirit and will vote persons into office whom he believes best qualified.

Disruption of a colony is possible through "poor" leaders and through "bad" leaders. Poor leaders lack native ability and training sufficient to the task; bad leaders lack integrity and use colony resources or their position for personal or family gain. Rivalry for leadership between the preacher and the householder are possible, but through long experience the chronicle has defined very clearly the role and relationship of each. Householders especially are warned against the temptations of selfishness, the desire to dominate, and the tendency toward laziness. Not only is the householder to be subject to the church and the preacher, but all of the varied stewards are to guard their work and resources carefully. In modern times it is unlikely, but possible, for a householder to squander the money of the colony through excessive drinking, even gambling, and unwise investments made without the consent of the colony. Any such offenses are reviewed and punished by the preacher assembly. On the informal level, a householder is often the son or brother of the preacher, and any tendencies toward rivalry or extortion are minimal.

There is little a colony can do to remove a "weak" leader if he does not commit a flagrant transgression. When a "bad" leader arises, there is invariably a factionalization of the colony into dominant family lines that vie for power. Often the early elements of this factionalization have culminated in the selection of a "bad" leader. If a colony is already factionalized into competing families, there is little a strong, capable leader can do to correct the situation other than to attempt to lead with a strong hand and to plan for an early branching. If an incapable leader is selected, the total structure of a healthy colony may carry it through this period of weak leadership. Similarly, most cohesive colonies can survive the tenure of a poor householder if the other members of the council are capable leaders. In some cases, the preacher assembly will step in to mitigate the situation. This is supportive and remedial.

Informal power struggles between groups of families can lead to disruption. When formal power is wrested from the colony by the nuclear family, or a group of families, usually a group of grown brothers, there can be serious cleavages. Colonies will rank each other by such factors as wealth, degrees of mechanization, managerial ability, amount and kind of defection by individuals, reputation of the preacher, and the degree of conformity to rules. Variations in these characteristics are to be expected. Some colonies that appear to be cohesive and well-managed to the outsider may actually be suffering from chronic internal troubles. Rarely have colonies resorted to courts of law to settle internal conflicts. Yet, in the case of cleavages along family lines, several suits over property ownership have been brought to the courts. Usually they occur among factions who have refused the advice of the judicial preacher assembly. In one unusual case, four brothers and their families, of a total of seven families in a colony, accepted a Protestant faith. The brothers refused to leave the colony, and asked the court to dissolve the colony corporation and distribute the assets among the seven families. The court ruled against the brothers.

Ideally, each nuclear family is subordinate to the welfare of the colony. In practice, a colony may be polarized between two or more sets of older brothers who hold important offices. Their mothers, more than their wives, are important in the informal leadership pattern. To maintain balance on the informal level, the role of women is crucial, for they must be socialized to "accept their place." Since women are believed to be inferior to men physically, intellectually, and emotionally, they must obey and follow their husbands. Through marriage, women unite (or form a link between) different family lines. This introduces an extended set of loyalties into the husband's family, which checks the exclusiveness of blood-brother relationships. The woman is more oriented toward her family of procreation than to her family of orientation although she maintains a loyalty to and a relationship with her parents and her sisters and brothers. She supports her husband in his competition with her brothers and with his brothers, and she always supports her grown sons. The man tends to maintain a deeper loyalty to his family of orientation than to his family of procreation. The differing loyalties of the husband and wife function to prevent the nuclear family from becoming too strong. Any agression the wife may feel against her husband for his seeming lack of emotional support is channeled against men in general and helps maintain a division between the male subculture and female subculture that functions to minimize sexual intrigue in a society that lives in close physical proximity. The woman's loyalty to the woman's subculture also offsets the emotional importance of the nuclear family in much the same way as the man's identification with the colony as a whole supercedes and thus weakens his personal identification with his wife and children.

Women have no voting privileges and no institutionalized means for acquiring formal leadership positions. Therefore, when women become powerful in intracolony politics, it is usually through an emphasis on the consanguinal family rather than on the good of the colony as a whole. When a preacher or householder is said to be strict, it generally means that he is strict in keeping women in their place. Within the present Hutterite culture this contributes to the smooth functioning of the colony as a unit, for it protects the society from self-centered, competing, family groups.

Defection

Several types of defection from the colony are disruptive to the social structure. There are individuals (usually young, unbaptized men) who "try the world" for several months or weeks, perhaps by following the harvest, but who return to colony life. There are a relatively small number of these "tourists" to the outside world who desert the colony permanently. Another type of defection is characterized by a group of "discontents" who defect from the traditional faith and give allegiance to some competing "renewal" group.

Since their migration to North America, relatively few persons reacted against the Hutterite system by abandoning it completely. The tendency for defections to increase in recent years does, however, give the elders concern. Only 258 men and 11 women left the colony voluntarily from 1880 to 1951 according to Eaton and Weil (1955:146), but over half returned; 106 men and 8 women were permanent defectors. Among the *Schmiedeleut* to 1960, 98 men and 7 women had defected (Mange 1963:32). An extensive study of factors associated with defection made by the authors leads them to conclude that the socialization patterns are so effective that the small number who do abandon the colonies are those who were disadvantaged or deprived of the normal training given a child.

The number of people who abandon colony life is highly correlated with a "declining" (as opposed to a cohesive) colony and can be accounted for in a few sibships. A declining colony (or as the Hutterites call it, *ein sterbene Kolonie*), has chronic problems of internal leadership that have not been resolved. The oldest colonies appear to have more structural problems than the newly established colonies. The mobility pattern does not permit a member to move from his home colony to another colony. Thus persons may not abandon a declining colony to join a more desirable one. Many of those who defected permanently had experienced long-term disputes with the householder, the preacher, or the German teacher. Leaving the colony in most cases was premeditated and touched off by arguments or antagonisms with superiors.

The Hutterites believe that defection occurs largely in certain "weak" family lines. This view of defection among themselves is valid, for of 38 defectors interviewed it was found that half originated in five families and all were located in declining colonies. The high status accorded to some families (usually those in leadership positions) and low status accorded to others were marked in these colonies. Nepotism was common in a declining colony and discrimination was invariably felt by the low-status individuals and families. The parents of many of those who abandoned the colony were atypical in some way: The relationship of the sons to their father was abnormal—either the father was very authoritarian or he showed favoritism toward one son. Out of a group of thirty-eight defectors, eleven were the sons of either preachers or householders. In a family where the father had been deposed as householder, seven children had defected from the colony.

Hidden competition for roles is normal, but it can become abnormal when one family represented on the council has too great an influence. There is some tendency toward primogeniture and inheritance of roles because a son, usually the eldest, acquires an intimate knowledge of the work of his father. The tendency toward role inheritance is functional. The cattle man, for example, must have the knowledge and ability to compete with the market conditions outside of the colony. It is important that the son know everything his father knew about cattle, and he must improve upon this knowledge. Strained relations between father and son can thus diminish the economic efficiency of the colony. Discontent is enhanced if the traditional roles are not equitably distributed among sibships.

When an adult or a leader flirts with alien values, the effect on the younger ones will be pronounced. The young, unbaptized man in a declining colony gets little help from his peers if he has ambitions of restoring order and the earlier idealism of his colony. Since adolescents are believed to be naturally "young and unproven," there is a particular problem in such a colony to maintain the traditional respect for authority. The discrepancy that develops between belief and practice will, for example, allow a person to have a radio, but he must not be caught listening to it.

Daughters have very different role strains than do sons. They share the subordinate status placed on women. If a family has a good reputation, the daughter has a good chance of an early marriage and does not feel the need to be pre-eminent or to rebel. Even though few women leave the colony, there is evidence that daughters of leaders do face atypical problems. All colony roles are psychologically marked by strong elements of dependency, but especially so for women. Any woman who cannot accept the definition of the role given to her will have more difficulty than a man in finding alternative avenues of expression. Several groups of sisters who abandoned a colony gave as their reasons that the colony was no longer Christian. An underlying factor is that these sisters, who found company in each other's misery, rejected the submissive Hutterite role assigned to women. Their conversion to an individualistically oriented fundamentalist denomination permitted a rational way of escape.

Overemphasis on the individual constitutes a disruptive influence in a communal society. It is rare that an individual will receive sufficient personal attention so that he will develop adequate personal security to leave the protective environment of the colony. Parents who show favoritism to a child or entertain ambitions for a child beyond those sanctioned by the colony introduce dissident elements and increase the probability of defection. This relationship is vaguely understood by Hutterites who say that if one is too good to a single child he is more likely to leave. The favored child who obtains attention and privilege above others in the family or in the colony, acquires self-confidence above his peers. Interests are developed and needs are felt beyond what the colony can provide. A child that does not experience the same rejection as his peers will not be frightened by problems that require imagination and individual solution. A father who wanted his son to become an engineer (a goal not attained by others in the colony) entertained noncolony goals for him. A father who wanted his son to become the English school teacher on the colony entertained a vocation for him that led to defection. An able leader who exceeds the limits (usually intellectually) or engages in certain privileges may unwaringly pave the way for subsequent deviation by his children. When children are treated as separate personalities in the formative years, individualism tends to develop to such an extent that it constitutes a threat to the colony. The system works best when the pre-eminence of the collective welfare over individual welfare is cheerfully and unquestionably accepted by all the members.

With few exceptions, the defectors interviewed were single when they left the colony, and found homes for themselves in predominantly rural areas. They found work similar to that performed in the colony. Very few moved directly into urban centers. Their present occupations are indicators of a lower-class style of life. Most of the defectors have become farm laborers rather than owners. Several are welders, a skill they acquired on the colony. Most are highly motivated in their job performance. The male defectors tended to be inactive or indifferent to religion. They showed little interest in making distinctions among the vast number of denominations that are not Hutterite. The absolutist position of the Hutterite religion continues to structure the thinking patterns of those who defect permanently. The religion taught to them in childhood is basically respected by them in such statements as: "If I ever want religion, I know where to find it." One who had abandoned the colony forty years ago said that during the intervening years, "the little faith that you had was right there to sort of guide you. The guardian angel was always with you, it seemed like."

When small groups of colony people defect from the colony there is great concern on all levels of leadership. Such instances have occurred in the past three decades when other communal groups have attempted to affiliate with the Hutterites. Most disruptive from the viewpoint of the Hutterites was their encounter with the Society of Brothers and the Community Farm of the Brethren. During most of their period in North America, Hutterites have had little interaction with other communal societies. The Harmony Society of Pennsylvania and the Amana Society of Iowa assisted the Hutterites financially in their early period, and the colonies maintained a cordial but distant relationship. In religious faith the Hutterites considered both of them too worldly. Hutterites have had occasional informal contacts with the Old Order Amish, and during the depression Amish congregations gave financial assistance to a needy Montana colony.

The founder of the Society of Brothers, Eberhard Arnold, was a German theologian who, as a result of deepening conviction, formed a small movement in Germany in 1922 devoted to brotherhood and communal living (Arnold 1964). His ideals were modeled after the early Anabaptists. After learning that Hutterites still existed in North America, he visited their colonies in 1930. Arnold made a profound impression on the colony leaders. His small colony in Germany was formally accepted into the Hutterite brotherhood, and he himself was ordained an elder by the Hutterites. Arnold died in 1935 and his followers moved from Germany to England and to Paraguay. Their hardships in Paraguay required the constant support of the North American Hutterites who generously contributed money, clothing, machinery, hardware, and tools. Although the Paraguayan groups have now disbanded, a daughter community exists in England and several in the United States.

The "Arnold people" wanted to restore the "missionary motive" within the traditional Hutterite institutions. The Hutterites acknowledged that they had lost their sixteenth-century practice of sending out missionaries. Reluctantly, a North American Hutterite colony accepted a few of the Paraguayan members into their colony. The ethnic Hutterites admired the learning of their new brethren, but discovered that they were unaccustomed to routine farm work. They preferred writing and discussions, painting, and enjoying nature in the twilight to operating a tractor or combine. They assumed leadership positions in the colony and a vast debt was accumulated in the name of the local colony. Dissention followed, and the group led by the Paraguayans decided to move to Pennsylvania, taking with them some of the ethnic Hutterites who sympathized with a less traditional view of religious practices, of child rearing, and of the participation of women in colony decisions. The Hutterites were very unhappy with the loss of their kin to the "Arnold people." When it was learned that the Society of Brothers tolerated smoking, folk dancing, musical instruments, divorce, and woman suffrage, it was not long until the Hutterite leaders severed formal connections with the Society of Brothers.

A similar encounter occurred with the communal attempts of Julius Kubassek, founder of the Community Farm of the Brethren. Kubassek, a Hungarian immigrant in British Columbia, affiliated with an Alberta colony. He was greatly impressed with the colony life and, with the full support of the Hutterites, formed a new colony near Bright, Ontario, in 1940. His colony attracted a variety of new converts and was also made up of some of the ethnic Hutterites who had joined the group. Factionalism developed later when it became evident to the Hutterites that Julius was too dictatorial, too individualistic, and also had tendencies toward celibacy. Although some of the Hutterite members returned, not all could do so, for several marriages had taken place. Two of the Alberta husbands refused to join their families who moved to Ontario.

In their history, Hutterites have lived under a variety of totalitarian and democratic governments, but never without interference. Although wars caused great hardship to their colonies, they managed to survive and to rebuild their communal pattern. Two periods of prosperity stand out in the four centuries of their history. The first was in the sixteenth century, marked by evangelism and growth by addition of new converts; the second was in the twentieth century marked by expansion and biological growth. Although there have been some legal restrictions affecting land purchases and occasional outbursts of hostility in times of war, disruption of colony life has been minor in North America in comparison with their earlier period.

6

The Genius of the Culture

THIS CASE STUDY has focused on those aspects of the culture central to Hutterite society. The background of the culture and its location in time was discussed in the introduction. A chapter each was devoted to world view, colony life style, technology and economic practices, socialization and family, and disruptive patterns. In this final chapter, the genius of the culture will be assessed. By genius is meant the native ability of the society to survive, to adapt to its environment, to meet the problems created by its own culture, and to provide for the basic human needs of the individuals in the community.

The culture demonstrates a unique social adaptation to a high rate of natural increase. The population characteristics of Hutterite society resemble that of a stable population model (Eaton and Mayer, 1954). The age-sex distribution within the group has remained fairly constant during its entire period in North America. The birth rate and death rates remain almost constant. The population remains overwhelmingly youthful, but not in the usual "primitive way" characterized by a high birth rate and a high infant death rate. The proportion of young children in the total population remains high due to a consistently high birth rate. Adults have about the normal life expectancy, but old people are a small minority because of the successive waves of new children. Thus, the fertility rate resembles that of underdeveloped countries, but the mortality pattern resembles that of industrialized countries. The high rate of natural increase requires appropriate social organization, flexibility, and expansion to meet these conditions.

The Hutterite society approximates the conditions required of any society to maintain a continuous high rate of population growth. These conditions include: (1) a stable society which is highly structured, (2) resources and factors favorable to expansion, and (3) a value system prescribing a high fertility rate (Lorimer 1954). The high fertility rate creates social problems which can be solved only through access to sufficient resources; in this case the acquisition of land and the accumulation of capital. Capital is accumulated by work and modes of behavior that are severely patterned. The values support a high fertility pattern and at the same time engender motivations that include sufficient spiritual and material rewards.

Typically most sects and subcultures are forced to change their social structure as the population expands. With the coming of industrialization and Western influence, many aboriginal bands of people in North America simply fled or disintegrated because they were unwilling to adapt their economy to the requirements of agriculture and a sedentary life. The Hutterites have not only fully accepted mechanized agriculture, but have adapted to a high rate of population increase by their institution of branching, which allows new units of the society to develop. The most significant social unit, the colony, is firmly controlled in size and location in a

Hutterites are rarely alone and seldom lonely.

manner that prevents urbanism and merchandising. Each colony is a primary social unit consisting of several nuclear families in a face-to-face group.

Hutterite society exhibits a remarkably stable pattern of communal living. Of all communal groups in the United States, the Hutterites are the largest, have the longest history, and are the most successful in maintaining communal life. There have been many attempts at communal living in the United States. Most of these have been short lived. Of 130 communal type settlements, 91 lasted less than a decade, 59 less than 5 years, 50 only 2 years, and 32 only 1 year (Deets 1940:22). Four endured for a century or more and are now extinct. They were the Amana Society, the Ephrata Cloister, the Shakers, and Old Economy. The Doukhobors, who organized in Russia in the eighteenth century, settled in western Canada in 1879 and are still living in several cooperative communities. When the Hutterites came to the United States in 1874, there were 8 different communal societies in 72 separate colonies. During the ensuing years, all 8 societies have declined or completely vanished as social systems while the Hutterites have grown from 3 to over 170 colonies.

The explanation of the stable social pattern is multifaceted. The goal of the society to live communally is achieved through a combination of factors. The most pervasive of these are acceptance of supernatural authority, a dualistic view of nature sanctioning separation from the world with a tendency toward isolation, limited addition of new members other than offspring, and a willingness to die or be killed rather than change their basic beliefs and social institutions. Supernatural authority is incarnated in the colony authority pattern. The belief that divine order governs material and social relationships is learned in infancy, perpetuated throughout life, and acknowledged on the death bed of the aged. The ordered universe is conceived as hierarchial, with one part always submissive to the other. The lower obeys and serves the higher. The central beliefs are unusually explicit about goals, subgoals, and norms of practice, and have a high degree of internal consistency. There is no speculation about theology, and logical thought and reasoned solutions are applied mainly to the economic activities of making a living. The basic ideology is supported by appropriate ritual, constant teaching, highly patterned social relationships, and clear delineation between power that may be democratically managed by man and power that is exercised only by God. The central beliefs appear to be sufficiently comprehensive to satisfy individual inquisitiveness within the scope of the colony environment.

The dualistic view of human nature, spiritual and carnal, is pervasive in the social organization and becomes the reasoned basis for separation from the world. The activities of the colony, the training of the young, and the management of specific problems are approached by the predetermined spiritual, rather than carnal, assumptions. The deification of social unity on a subverbal level and the submission of the self to the will of God tends to minimize arguments and speculations of a theological nature. Geographic isolation as well as symbolic isolation through language and dress are manifestations of the central belief in separation from the world. Isolation effectively minimizes the problems of communal living as the Hutterites practice it.

The small number of nonethnic converts to the colonies is conducive to a stable social pattern. The practice of sending out missionaries to preach and make converts outside of the community in their early period of history has been discontinued entirely. This practice eliminates the problems of assimilating large numbers of outsiders. As the Hutterite movement became institutionalized and strongly patterned, the group relied on its own offspring for new members. In its charismatic or prophetic stage the movement attracted converts, but during the seventeenth century the group tended toward institutionalization.

Hutterite child rearing and socialization practices are phenomenally successful in preparing the individual for communal life. The individual is taught to be obedient, submissive, and dependent upon human support and contact. Socialization is consistent and continuous in all age groups. From early childhood to adulthood there is no relaxation of indoctrination within clearly defined age and sex groupings. Every individual is subservient to the colony at every stage of his life. The goals for each stage of socialization are attainable by virtually all Hutterites. Individuals are well trained to meet clearly defined roles, and each is rewarded by the smooth execution of his work and by the awareness that his contribution is needed by the colony. A certain amount of deviance of an acceptable type is permitted within each age set and relieves the system of stifling rigidity.

Socialization is achieved in a supernaturally based authority system. All elements of knowledge, motivation, and activity are directly related to a single source of authority and are subordinated to it. No segment of knowledge remains unrelated to the dominant source of authority nor is any permitted to develop aside from it. The scope of socialization includes a conception of time that is temporal as well as eternal and involves the active participation of the individual in both. Ritualization of temporal and sacred activity is well proportioned. The individual is socialized within a spatial relationship that includes all of the social institutions, activities, and resources he will need as an individual. Interaction with and dependency upon outsiders is minimized.

In infancy the child learns to enjoy people and to respond positively to many persons. After age 3 he is weaned away from his nuclear family and learns to accept authority in virtually any form. He learns that aloneness is associated with unpleasant experiences and that being with others is rewarded with pleasant experiences. During the school years he is further weaned from his family, learns more about authority, and acquires a verbal knowledge of his religion. He acquires the ability to relate positively to his peer group and to respond to its demands. As a small child his universe is unpredictable, but as he matures in his peer group and takes part in colony life his universe becomes highly predictable. He learns to minimize self-assertion and self-confidence and to establish dependence on the group. As a member of a categorically defined age set the young child learns explicitly when and whom to obey. School-age children receive limited companionship and little indulgence from their parents; they learn how their peer group protects and punishes them; they learn to accept frustrations passively, and to enjoy hard physical labor. When the child becomes an adult, he is rewarded with responsibility, privileges, and greater recognition and acceptance by his nuclear family. After becoming a full member of the colony through baptism, the self-image expands to include the whole colony with which he identifies.

The expectations for the individual within the social structure is clearly defined by the ideology and is reinforced by the social patterns. Each individual knows what is expected of him; he wants to follow these expectations and, in most instances, is able to do so. He identifies ideologically and emotionally with the colony system. There is a strong aversion to the ways of the *Draussiger,* the outsider, who is a child of the world. The Hutterite looks at himself as belonging, not to a world created by Newton, Beethoven, Sartre, or Einstein, but to the model described in the Bible. The colony is for him, as for certain other ascetic Christian groups, a paradise surrounded by vast numbers of unconverted human beings whose destiny is determined by God and whom he will not judge.

The ascetic demand of the society does not ask the individual to mortify the basic human drives, but to subject them to a community of love that is both human and divine. Sex is managed in such a way that it is not a threat to the reproductive patterns, as with some communal Christian groups. In its socially sanctioned place, the practice of sex is distinguished from the "lust of the eye," which is illegitimate desire. Sex is limited to married pairs, and there is no evidence of any tendency toward celibacy or continence after marriage. Families are generally happy and supportive of the colony's superior power position. As an individual, a Hutterite may demonstrate what appears to the outside as egocentrism or pride. But to the Hutterite who knows that his system is right, he is not expressing personal pride but simply a state of mind that has no doubts. The emotional identification with the colony is evidenced by the number who return after "trying the world" and the relatively few who permanently desert the colony.

The Hutterite appraisal of human nature is functional in the communal system. In some respects it is dynamic. A child's nature is regarded as carnal (selfish) and must be supplanted by the spiritual (selfless) nature. The spiritual nature is expressed within a social hierarchy that is based on moral obligations among the members. The Hutterites regard the child's personality as intrinsically good only as he (behaviorally) gives up his individual will and conforms voluntarily to the will of the colony. The consequences of this appraisal of human nature for delinquent individuals is somewhat opposed to that of contemporary society in North America. When a Hutterite youth transgresses the rules, he will be punished, but not because he was bad. Since the society knows everyone is bad from the start, it is required that the wrongdoer suffer the consequences of his behavior. The society expects that the badness will, with maturity, be supplanted by responsibility. The adolescent who occasionally transgresses the rules can still see himself as being on the road toward goodness. Hutterite society makes it a point to forgive when true signs of betterment are observed. These manifestations are encouraged and rewarded. When the individual becomes a full member, he is obligated to teach and uphold the morality of his society.

In North American society, by contrast, the child is thought of as good, primarily needing a chance for self-development and self-realization. When the child transgresses the rules of the society, he is disappointing. The adolescent who comes into conflict with the law tends to be labelled as bad and this will stand against his reputation in the future. Transgression has a certain irreversible and unredemptive direction, and the road to goodness becomes almost exclusively the task of the individual himself. The Hutterite practice tends toward the acceptance of deviancy within manageable proportions, frequently within a small peer group. The Hutterite position is simply that the individual is not perfect and cannot attain perfection without the aid of his brothers. The net effect is a view of the individual that leads toward rehabilitation with a minimum of condemnation. The Hutterite individual knows why he is alive, how he should live, and subjectively he feels able to meet the standards of behavior.

Innovation without acculturation: innovation is selective, managed, and tends to be consistent from the viewpoint of the dominant goal of communal living. Innovation does not alter the internal hierarchy and social structure. As in all societies, culture change among Hutterites is selective and proceeds according to a pattern that is more or less consistent with their world view. Innovations are evaluated with reference to the basic objectives of communal living. Hutterite society is unlike the ideal, typical folk society in several important respects. The ideology of intentional communal living contrasts sharply with the nonliterate style of primitives. Their rational-bureaucratic approach to economic and technological aspects of living contrasts sharply with the tradition-directed type of society. Their native ability to adapt to change is important to their survival in the modern world. The past migrations of the group with their strong preference for living in countries and in regions geographically isolated from large urban centers have minimized many of the complex problems brought about by change. When the external pressures became too great in the past, the group has moved to other lands.

Innovations in the colonies are carefully considered and evaluated on the basis of their possible effect on communal living. Informal talk always precedes any formal discussion of change. Modernity is not necessarily stereotyped as being evil or worldly. Most leaders fail to acknowledge any conflict between religious beliefs and technological change. "Nothing is too modern if it is profitable for the colony." The dominant criterion for introducing change is whether the innovation supports the welfare of the colony. A firm boundary is drawn between changes that involve the economic well-being of the whole colony and changes that are primarily for personal comfort and convenience. "We do not believe in making everything nice for the flesh," they say. Installing showers (in addition to tubs that they already had) in the central laundry required more time and discussion in one colony than buying a new combine. Changes are considered safe when they are anchored on spiritual rather than carnal grounds. Innovations in the outside world, by contrast, are believed to be made on the basis of satisfying the carnal nature of man. Changes are spaced, as too many changes at one time are not considered good for the colony. Even more important for survival than "controlled acculturation" (Eaton 1952) is the necessity that all members share alike in any changes permitted. In this way the relative status of any one individual to others has not changed at all. No person is made richer or poorer or more powerful over others. The preacher is still the spiritual leader, the younger are still subordinate to the older, and women are still subject to rule by men. The use of trucks, tractors, mechanized poultry and livestock raising in the colony, and the introduction of sewing machines and electric irons have not changed the social relationships. The Hutterites adjust to a changing technology, but they would rather die than change their basic social patterns, which they believe to be ordered by supernatural authority.

Hutterites maximize the power of informal, primary group consensus. The daily, constant association between leaders and followers, among persons of all ages, tends toward solidarity and an intimate knowledge of the values and opinions held by other persons. Normative practices tend to develop and determine the issues that will be discussed formally. Nothing is discussed formally that is not already a problem or an issue of concern to the group. When there is pressure for change on the part of the foreman and managers, the leaders are able to assess it informally, and after there is widespread informal agreement, the colony formally votes on an issue. Decisions are made on the basis of a majority vote. The acceptance of innovations by consensus before there is widespread breaking of the rules and before pressures become unmanageable has the effect of maintaining members' respect for the authority system. The emphasis on group welfare and de-emphasis on individual rights and convictions minimizes the friction often associated with innovation in sectarian societies.

On the formal level, individual colonies may allow only those changes that are in keeping with the preacher assembly. A given colony is not autonomous but is subject to the discipline (*Gemeindeordnungen*) of the preacher assembly. These rules perpetuate uniformity among colonies, and slow down the rate of innovation. The balance in decision making between any given colony and its preacher assembly appears to have great advantages in terms of satisfying human needs. When one or more colonies deviate from a formal disciplinary rule, the preacher assembly will place the issue on their agenda. The violation may be discussed and the rule that was violated may be reaffirmed by majority vote; it may be modified, or a vote may be postponed to allow time for further informal consensus. Issues of special concern to the preacher assembly result in a formal statement, which is circulated and read to all colonies by their preacher. The colonies in turn vote whether to accept the ruling. Thus, while the Hutterites give their complete loyalty to a theocratic rule, the rules that govern behavior are democratically managed. The reason for changes are neither individual wants nor expediency, but are based entirely on the welfare of the colony.

Cleavages that normally develop between competing families or individuals for acquiring property in North American society have no formal basis for existence in Hutterite society. Consequently, material innovations are managed so as not to be disruptive. Family units are not competing against each other for wealth. When a colony is prosperous, the family units benefit equally, and none are symbolically excluded. Wealth, property, and knowledge are denied as a means of social distinction. The competition that is often very marked between colonies is integrative rather than disruptive to any given colony. Any impression, however, that the colonies express no anxiety about the effects of change would be misleading. The contacts engendered by commerce and trade with the large urban centers near their colonies have undesirable consequences for them.

Through its many self-sustaining colony units, Hutterite society is sufficiently patterned to integrate the meaning of economic values with spiritual values. Each colony is a believing community, a community of worship, and a community of work. The absolutism of their ideology has made it necessary for them to migrate to various countries where they have tended to live in isolation from their neighbors. The ideology has not forced the members to isolate themselves from biologic, economic, and technological solutions to living. The group has retained vital links between the generations. Hutterite society is a school, and the school is a society. In shaping patterns within the culture, the group has managed the problem of fragmentation, of individualistic and dictatorial leaders, defection, and senile decay. The bond of community is defined in such a way as to include the physical and emotional components of human need. The deep feelings for the correctness of their way of life was expressed by a thoughtful Hutterite who said: "If there will ever be a perfect culture it may not be exactly like the Hutterites—but it will be similar."

References Cited

ARNOLD, EBERHARD, 1964, *Eberhard Arnold, A Testimony of Church-Community from His Life and Writings.* Rifton, N.Y.: The Plough Publishing House.

*BENNETT, JOHN W., 1967. *The Hutterian Brethren: Agriculture and Social Organization in a Communal Society.* Stanford, Calif.: Stanford University Press.

CANADIAN MENTAL HEALTH ASSOCIATION, 1953, *The Hutterites and Saskatchewan: A Study of Inter-group Relations.* Regina, Saskatchewan.

*CLARK, BERTHA W., 1924, "The Hutterian Communities," *Journal of Political Economy,* Vol. 32, pp. 357–374 and 468–486.

*CONKIN, PAUL K., 1964, *Two Paths to Utopia: The Hutterites and the Llano Colony.* Lincoln, Neb.: The University of Nebraska Press.

*DEETS, LEE EMERSON, 1939, *The Hutterites: A Study in Social Cohesion.* Gettysburg, Pa.: Times and News Publishing Co. Ph.D. dissertation, Faculty of Political Science, Columbia University.

EATON, JOSEPH W., 1952, "Controlled Acculturation," *American Sociological Review,* June, pp. 331–340.

———, 1964, "The Art of Aging and Dying," *The Gerontologist,* Vol. 4, No. 2.

EATON, JOSEPH W. and ALBERT J. MAYER, 1954, *Man's Capacity to Reproduce: The Demography of a Unique Population.* New York: Free Press. Reprinted from *Human Biology,* Vol. 25, No. 3.

*EATON, JOSEPH W. and ROBERT J. WEIL, 1955, *Culture and Mental Disorders: A Comparative Study of Hutterites and Other Populations.* New York: Free Press.

ELIADE, MIRCEA, 1961, *The Sacred and the Profane.* New York: Harper & Row, Harper Torchbook edition.

*FRIEDMANN, ROBERT, 1961, *Hutterite Studies.* Goshen, Ind.: Mennonite Historical Society.

———, 1965, *Die Schriften der Huterischen Täufergemeinschaften. Gesamtkatalog Ihrer Manuskriptbücher, Ihrer Schreiber und Ihrer Literatur* 1529–1667. Wein: Hermann Böhlaus Nachf.

*GROSS, PAUL S., 1965, *The Hutterite Way.* Saskatoon, Saskatchewan: Freeman Publishing Co.

HOEBEL, E. ADAMSON, 1966, *Anthropology: The Study of Man,* 3d ed. New York: McGraw-Hill.

HOFER, PETER, 1955, *The Hutterian Brethren and Their Beliefs.* Starbuck, Manitoba:

---

* Recommended for additional reading and for advanced term papers.

The Hutterian Brethren of Manitoba. Approved by the Committee of Elders.

*HORSCH, JOHN, 1931, *The Hutterite Brethren 1528–1931: A Story of Martyrdom and Loyalty.* Goshen Ind.: Mennonite Historical Society.

HOSTETLER, JOHN A., 1963, *Amish Society.* Baltimore, Md.: Johns Hopkins Press.

———, 1965, *Education and Marginality in the Communal Society of the Hutterites.* University Park, Pa.: The Pennsylvania State University. Mimeographed.

HUTTERIAN BRETHREN CHURCH, 1950, "Constitution of the Hutterian Brethren Church and Rules as to Community of Property." Winnipeg, Manitoba: E. A. Fletcher, Barrister-Solicitor.

HUTTERISCHEN BRÜDERN IN KANADA, 1961, *Gesang-Büchlein. Lieder Für Schule und häuslichen Gebrauch.* Cayley, Alberta.

———, 1962 *Die Lieder der Hutterischen Brüder.* Cayley, Alberta.

KAPLAN, BERT, and THOMAS F. A. PLAUT, 1956, *Personality in a Communal Society.* Lawrence, Kan.: University of Kansas Publications.

KLASSEN, PETER J., 1964, *The Economics of Anabaptism 1525–1560.* The Hague: Mouton and Co.

KLUCKHOHN, CLYDE, 1949, "The Philosophy of the Navaho Indians," in *Ideological Differences and World Order,* ed. F. S. C. Northrop, New Haven, Conn.: Yale University Press, pp. 356–383.

LITTELL, FRANKLIN H., 1958, *The Anabaptist View of the Church,* 2d ed. Boston: Starr King Press.

LORIMER, FRANK, and others, 1954, *Culture and Human Fertility.* Paris: UNESCO.

MANGE, ARTHUR P., 1963, *The Population Structure of a Human Isolate.* Ph.D. dissertation, Department of Genetics, University of Wisconsin.

NORDHOFF, CHARLES, 1961, *The Communistic Societies of the United States.* New York: Hillary House Publishers. Original ed., 1875.

OBERNBERGER, ALFRED, 1966, *Die Mundart der Hutterischen Brüder.* Unpublished manuscript.

PETER, KARL, 1965, *Social Class and the Conception of the Calling. Toward a Constructive Revision of Max Weber's Hypothesis.* Master's thesis, Department of Sociology, University of Alberta, Edmonton.

*PETERS, VICTOR, 1965, *All Things Common: The Hutterian Way of Life.* Minneapolis: University of Minnesota Press.

PRIESTLY, DAVID T., 1959, *A Study of Selected Factors Related to Attitudes Toward the Hutterites of South Dakota.* Master's thesis, South Dakota State College.

RIDEMAN, PETER, 1950, *Account of Our Religion, Doctrine and Faith.* Bungay, Suffolk, England: Hodder and Stoughton, Ltd., in conjunction with the Plough Publishing House. Trans. by K. E. Hasenberg from the German edition of 1565.

RILEY, MARVIN P., and DAVID T. PRIESTLY, 1959, "Agriculture on South Dakota's Communal Farms," *South Dakota Farm and Home Research,* Vol. 10, No. 2, Agricultural Experiment Station, South Dakota State College.

SAHLINS, MARSHALL D., 1965, "On the Sociology of Primitive Exchange," in *The Relevance of Models for Social Anthropology.* London: Tavistock Publications; New York: Praeger.

SERL, VERNON C., 1964, *Stability and Change in Hutterite Society.* Ph.D. dissertation, University of Oregon.

SMITH, C. HENRY, 1957, *The Story of the Mennonites.* Newton, Kans.: Mennonite Publication Office.

WEBER, MAX, 1930, *The Protestant Ethic and the Spirit of Capitalism,* trans. by Talcott Parsons. London: Allen and Unwin.

WILSON, BRYAN R., 1959, "An Analysis of Sect Development," *American Sociological Review,* Vol. 24, pp. 3–15.

WOLKAN, RUDOLF, 1923, *Geschicht-Buch der Hutterischen Brüder.* Wien, Austria, and Standoff Colony, Alberta.

ZIEGLSCHMID, A. J. F., ed., 1943, *Die älteste Chronik der Hutterischen Brüder*. Philadelphia: Carl Schurz Memorial Foundation.

———, 1947, *Das Klein-Geschichtsbuch der Hutterischen Brüder*. Philadelphia: Carl Schurz Memorial Foundation.

Recommended Film

The Hutterites. Educational film produced by the National Film Board of Canada, Montreal, Quebec. Black and white, 16mm, 28 minutes. Available in the United States for purchase through Sterling Educational Films, Inc., 241 East 34th St., New York, N.Y., and for rental through Contemporary Films, Inc., 267 West 25th St., New York, N.Y.; 614 Davis St., Evanston, Illinois; and 1211 Polk St., San Francisco, California.

052417